THE
TRUTH

THE RATIONALIST'S GUIDEBOOK
TO RAISING HUMANE SAPIENTS
IN A BIG BRAINED APE WORLD

A SOCRATIC PROGRESSION
OF SEQUENTIAL ESSAYS

The Humane Sapient
Enlightenment Foundation

My unending love and gratitude to my wonderful parents, who always told me The Truth, and immersed me in the Natural World early and often in my formative years, while teaching me their honorable ethics by their exemplary character and integrity. They are responsible for what is best in me.

ISBN: 9781484956427   Available at:
https://www.createspace.com/4281303 (Print)
http://www.amazon.com/dp/B00EY296XM (Kindle)
©MMXIII The Humane Sapient Enlightenment Foundation
A Non-Profit Educational Foundation. All Rights Reserved
4957 Lakemont Blvd SE, Suite C-4 #238, Bellevue, WA 98006
TheHumaneSapientEnlightenmentFoundation.org
DarwinLovedYou.com

Front cover watercolor portrait of Charles Darwin painted by George Richmond in the late 1830s after Darwin had returned from his five year circumnavigation of the globe aboard the HMS Beagle.

No part of this publication may be reproduced, stored in a retrieval system, or transmitted, in any form or by any means, electronic, mechanical, photocopying, recording, or otherwise, without the expressed written permission of the author.

No patent liability is assumed with respect to the use of any of the information contained herein. Although every precaution has been taken in the preparation of this book, the publisher and author assume no responsibility for errors or omissions. Neither is any liability assumed for any damages resulting from the use of any of the information contained herein.

"Take this brother, may it serve you well."
-John Lennon-

For Bean and DJ and
All Children Everywhere

CONTENTS 1

THE TRUTH... 15

TRUTH FROM LIES?... 17

IGNORANCE... 19

HERITAGE... 21

TRUTH IS NOT SUBJECTIVE... 23

THE SCIENTIFIC METHOD... 25

ORIGIN STORIES... 27

DARWIN LOVED YOU... 29

RELIGION... 31

ATHEISM... 33

DEATH... 35

ETHICS... 37

SONS OF ABRAHAM... 39

SONS OF GREAT BRITAIN... 41

FOUNDING FATHERS... 43

GOVERNMENT... 45

DNA... 47

RACE... 49

WAR... 51

THE COLD WAR... 53

SPACE COLONIZATION... 55

HUNGER... 57

EDUCATION... 59

CONTENTS 2

MOTHER EARTH... 61

WATER WARS... 63

SOLAR WISDOM... 65

NUCLEAR WISDOM... 67

SCIENCE AND POLITICS... 69

VACCINATION... 71

CIRCUMCISION... 73

BREAST FEEDING... 75

OWNER'S MANUAL... 77

A BBA AND THEIR MONEY... 79

WILD APES... 81

SUGAR... 83

OPTIMUM HEALTH... 85

BREAST CANCER... 87

NICOTINE... 89

THE BRAIN... 91

SEXUALITY... 93

ABORTION... 95

GAY MARRIAGE... 97

AIDS... 99

WWW... 101

TEENAGE DRIVERS... 103

FIREARMS... 105

CONTENTS 3

CRIME AND PUNISHMENT... 107

DUI... 109

POLICE... 111

PRISON... 113

CITY KIDS... 115

APE HIVES... 117

WEALTH... 119

THE STATE... 121

DECEPTION... 123

THE FED... 125

TAXES... 127

SELF- INFLICTED INJURY... 129

EXTINCTION... 131

FINAL THOUGHTS... 133

DISCOVER THE OBJECTIVE TRUTH... 135

AWAKE, RISE, EDUCATE AND ACT!... 137

EVIL APES OF AVARICE... 139

WHAT ARE YOU PREPARED TO DO?... 141

'CENTRAL BANKING' FOR BBA 'CHILDREN'... 143

TEACH YOUR CHILDREN WELL... 145

LISTEN LESS AND THINK MORE! LLTM!... 147

"If you are out to describe The Truth, leave elegance to the tailor."
"If you can't explain it simply, you don't understand it well enough."
-Albert Einstein-

NOTES

THE TRUTH

"Fear not the path of Truth for the lack of people walking on it." - Robert F. Kennedy

All essays must be read in their sequential order, as later essays will refer to, and assume a knowledge of, all earlier essays.

"There are two ways to be fooled. One is to believe what isn't True; the other is to refuse to believe what is True." - Soren Kierkegaard

As there is no genuine substitute for independent research, to facilitate better understanding and motivation to take action, it is the author's recommendation, that if at any time, the reader disputes any of the definitive assertions contained herein, they should stop reading and invest the time to *independently research and confirm* their objective Truth before continuing; as the author affirms that each and every assertion contained herein is self-evident, if one simply knows The Truth.

"All Truths are easy to understand once they are discovered; the point is to discover them."
"We cannot teach people anything; we can only help them discover it within themselves."
-Galileo Galilei-

NOTES

TRUTH FROM LIES?

"Beware of false knowledge; it is more dangerous than ignorance." - George Bernard Shaw

Would it ever be possible to arrive at a True conclusion, when starting from a False premise?

"I shall not mingle conjectures with certainties."
-Sir Isaac Newton-

NOTES

IGNORANCE

*"Whenever you find yourself on the side of the majority,
it's time to pause and reflect."* - Mark Twain

In an Age of universal ignorance, concealing The Truth is a simple matter.

By definition, the ignorant don't know they are.

Ignorance is never superior to knowing The Truth, and can even be fatal.

The Truth is free, it is ignorance that is costly.

"Being ignorant is not so much a shame, as being unwilling to learn." "We are all born ignorant, but one must work hard to remain stupid." - Benjamin Franklin

Question everything you've ever been told or shown,
and all you think you know; much of it is false.

And teach your children to always do the same.

"The unexamined life is not worth living." - Socrates

The best one can hope for in this brief singular life is to discover The Truth;
and know it when they do.

"The Truth of things is the chief nutriment of superior intellects."
-Leonardo da Vinci-

NOTES

HERITAGE

"There is no absurdity so palpable but that it may be firmly planted in the human head, if only you begin to inculcate it before the age of five, by constantly repeating it with an air of great solemnity." - Arthur Schopenhauer

Ignorance and superstition are passed from generation to generation.

Break the chain.

Don't allow pernicious ignorance to remain your family's heritage.

"The reason there's so much ignorance is that those who have it are so eager to share it." - Frank Howard Clark

Emotion proves to be the enemy of Truth.

"Above all, don't lie to yourself." - Fyodor Dostoevsky

The Universe is not limited to your ability to understand it.

"There are more things in Heaven and Earth, Horatio, than are dreamt of in your philosophy." - William Shakespeare

Never allow another's limitations to become your own.

And never pass your own limitations on to your children.

Free them from the delusions of their ancestors' false mythologies.

Teach your children The Truth; not Bronze Age *lies and nonsense*.

"Identity is partly heritage, partly upbringing, but mostly the choices you make in life."
-Patricia Briggs-

NOTES

TRUTH IS NOT SUBJECTIVE

"The Truth is incontrovertible, malice may attack it, ignorance may deride it, but in the end; there it is." - Winston Churchill

The Truth exists independent of our *'belief'* in it, or even our mere awareness of it. And nothing becomes True, simply because we *'believe'* it to be True.

"An error does not become Truth by reason of multiplied propagation, nor does Truth become error because nobody sees it." - Mohandas Gandhi

The Earth has always been a spherical rock hurtling through nondescript Space ever since it first coalesced out of the dust and ice of the protoplanetary disk of our rotating solar orbital plane four and a half billion years ago. It was never flat, nor the center of the Universe, nor even the center of our own Solar System; despite our Bronze Age ancestors all *'believing'* it to be. That *objective* Truth didn't change, only our own species' levels of *subjective* ignorance and awareness of it.

"A Truth's initial commotion is directly proportional to how deeply the lie was believed. It wasn't the world being round that agitated people, but that the world wasn't flat. When a well-packaged web of lies has been sold gradually to the masses over generations, The Truth will seem utterly preposterous, and its speaker a raving lunatic." - Dresden James

The objective Truth has existed for billions of years before we evolved, and will continue to exist for billions of years after we're extinct; wholly independent of our species' subjective awareness of it. The objective Truth never changes to *adapt* itself to our species' subjective ignorance, but rather our species must always change to adapt ourselves to newly discovered objective Truth. As the closer all of our individual, and species' collective subjective awareness matches the objective Truth, the more deeply enriched is the quality and longevity of our own individual lives, and the more likely is our species' collective thriving survival.

"Nature is relentless and unchangeable, and it is indifferent as to whether its hidden reasons and actions are understandable to man or not." - Galileo Galilei

But to understand The Truth, one must first discover it; as each Era is blind to its own darkness.

"If you believe the doctors, nothing is wholesome; if you believe the theologians, nothing is innocent; if you believe the military, nothing is safe." - Lord Salisbury

Teach your young children to always; Listen Less and Think More. (LLTM)

"If we wonder often, the gift of knowledge will come." - Arapaho Proverb

Teach them the daily habit of spending at least an hour in quiet solitude; away from any distracting 'media' of any kind; calming their inner dialog, exclusively contemplating the whole of the Natural World. For the objective Truth is there for them to instructively witness; they have but to let it in.

"Look deep into Nature, then you will understand everything better."
"Any man who reads too much, and uses his own brain too little, falls into lazy habits of thinking."
-Albert Einstein-

NOTES

THE SCIENTIFIC METHOD

"Science is the search for Truth; not a game in which one tries to beat his opponent..." - Linus Pauling

The Scientific Method is the only time tested, *evidence based*, proven process for conclusively confirming the objective Truth. Every modern convenience that we now take for granted is the result of applying The Scientific Method. It's no accident that all inventions of consequence were developed by Scientists from widely diverse cultures, strictly adhering to The Scientific Method. Not because of their own culture's unique Origin Story or politics; but often in spite of both.

*"In questions of Science, the authority of a thousand is not worth
the humble reasoning of a single individual."* - Galileo Galilei

The Scientific Method is the only experience proven, culturally neutral universal language. It's process of *peer review* eliminates both personal and cultural bias. If an objective third party can't duplicate your results, your erroneous conclusions are discarded. Science has no interest in what you may *'think'*, *'feel'* or *'believe'*; but only in what you can prove.

"A man should look for what is, and not for what he thinks should be." - Albert Einstein

Since the 1500s 'The Method' alone has uniquely assured the objective Truth and Intellectual Integrity. Teach your children to seek only that Truth which can be *proven*, and to avoid all of the ignorant and superstitious and deluded who waste their entire brief singular life arguing over what can't be known; for when the wise argue with fools, observers can't tell them apart.

"Never tell The Truth to people who are not worthy of it." - Mark Twain

Science can be intuitive; once one understands 'The Method' they can often quickly determine the probable validity of new claims based on the ever growing 'body' of our species' collective Empirical Knowledge from previous proven work. You don't have to be able to build a car, to be able to drive one. The only constant in our evidence based Empirical Knowledge is growth.

"If I have seen farther than others it is by standing on the shoulders of giants." - Sir Isaac Newton

This is the first ever generation in our entire history to have our collective Empirical Knowledge instantly available online anywhere on this planet. Encourage your children to always seek its Truth in every endeavor. Teach your children while young the disciplined Reasoning process of The Scientific Method to empower them throughout their lives with the only tool they will ever need to discern The Truth from the specious; so they may always make judicious life choices. The most important instruction you must impart to your children is *how* to think, rather than *what* to think; and to clearly distinguish the impossible, from the not yet imagined; to prepare them for their future life experience likely to eclipse our own in wonders not yet dreamed. We don't know what they will encounter, but their odds of success will be greatly improved come whatever may, with the ongoing application of our collective Empirical Knowledge and The Scientific Method. The global Scientific Community is less than one percent of the total population. They actually know *what we are,* and *where we came from,* and therefore; *where we should be headed.* Their greatest failure has been in not yet disseminating that knowledge to all the rest of the population, where pernicious subjective ignorance of our True origins remains the norm for the vast majority.

*"Give a man a Truth and he will think for a day.
Teach a man to Reason and he will think for a lifetime."*
-Phil Plait-

NOTES

ORIGIN STORIES

"The further a society drifts from Truth, the more it will hate those who speak it." - Selwyn Duke

Teach your children that Ethnologists have recorded hundreds of different Origin Stories from cultures all over this planet, and share many of those *imagined* stories with them; including the *'talking serpent'* among all the many fanciful anthropomorphic tales. Teach them that because all of those Origin Stories are contradictory, they are mutually exclusive; therefore only one, if any, can be True. Teach them that exclusively only one alone is supported by any evidence, and that one is supported by all the evidence ever obtained through the proven, objective, culturally neutral Scientific Method. And without a single contradiction of any kind; everything that we yet learn today in every Scientific discipline, only adds to that ever growing body of our evidence based Empirical Knowledge supporting that one Truth. That one objective Truth didn't change, only our own species' levels of subjective ignorance and awareness of it.

That Truth isn't going away, and exists without your awareness of it, but your children need and trust you to courageously teach them only proven Truth; as the quality of their lives depends on it. Do not fail them! Don't teach them imagined Origin Stories; teach your children *The One Proven Truth;* 'Human Beings' are Big Brained Apes (BBAs). We didn't evolve *from* Apes, we *are* Apes; mostly hairless Big Brained Apes.

"Anyone who challenges the prevailing orthodoxy finds himself silenced with surprising effectiveness. A genuinely unfashionable opinion is almost never given a fair hearing in the press." - George Orwell

Because only this one Origin Story is proven to be True, the others are all imagined delusive myths, contradicted by all evidence, which must now be abandoned if we are ever to advance our species' Moral Evolution to achieve lasting Peace and Justice on this planet. As our species' juvenile 'belief' in divisively irreconcilable, mutually exclusive Origin Story 'Holy Books'; the archaic Bronze Age factual equivalents of Harry Potter novels; is the underlying etiology of all BBA conflict on Earth.

No Darwin; No Peace. Know Darwin; Know Peace.

Our species' irrational propensity for 'belief' in things we only imagine may even be genetic; having been 'selected for' in us as the descendents of those Apes who survived in higher numbers by first skittishly fleeing suspected savanna predators; they had not yet confirmed, but only imagined were present; before their 'nonbeliever' cousins waited to confirm the danger before fleeing, but then only too late to escape becoming meals in much higher gene thinning numbers. That *'faith'* in the existence of something unseen and unconfirmed may therefore have 'selected' those Apes who only imagined a presence, of what didn't actually exist in reality, as a safe 'flight distance' survival strategy. Our many million year 'hard wiring' of which not only became unneeded many thousands of years ago after the better armed BBAs had exterminated most of the megafauna that preyed upon us; but has now even been endlessly demonstrated to be homicidally detrimental to our continued survival; as we replaced their predation with our own, as the only species capable of killing from a distance; and whom now uses that unique ability to significantly cull their own numbers due largely to our juvenile 'faiths' in divisively irreconcilable, mutually exclusive imagined Bronze Age Origin Story 'Holy Books'.

The hard wired genetic basis of all 'faith' is *fear;* of the dark, of the unknown, and of the imagined. Change in the environment will change its 'selective' pressures. Traits that were once beneficial may then become detrimental. Being afraid of the imagined originally helped us survive; but it may now induce our extinction. We must *adapt or perish.* Darwinism: where all your answers are questioned.

"There are three classes of people:
Those who see. Those who see when they are shown. Those who do not see."
-Leonardo da Vinci-

NOTES

DARWIN LOVED YOU

"We cannot judge The Truth of an idea by our fears of its effect." - Stefan Molyneux

Teach your children that Evolution is a Scientific Method conclusively proven objective Truth. What Charles R. Darwin brilliantly postulated in his 1859 book; *"On The Origin of Species by Means of Natural Selection..."* has now been conclusively confirmed through genetic research; 'Humans' are Big Brained Apes (BBAs). It takes great courage and humility to accept The Truth. The 'fossil record' and DNA do not lie; only cowardly Big Brained Apes do; even to themselves. To transcend the limitations of our Simian heritage we must all first courageously embrace them. All calendars will one day show his insight's 1859 publishing date as year one AD. (After Darwin) Excluding all the false Origin Stories' dates preceding it from the dark years BD. (Before Darwin) In an Age of universal deceit, telling The Truth is a noble act of love. OMD! (Oh My Darwin!)

"When a true genius appears in this world you may know him by this sign, that the dunces are all in confederacy against him." - Jonathan Swift

The simplest definition of Evolution Science is it's the study of diversification from a common ancestor. What the *'faithful'* don't understand about their 'denial' of The Truth of Evolution is that Evolution is the sole foundation and single unifying principle of all modern Scientific disciplines. From Astrophysics and Paleogeology, to Biochemistry and Genetic Medicine; without Evolution, there would be no 'modern Science'. Those 'faithful' all have an unethical tendency of selectively embracing only those 'scriptures' that support their own absurd assertions. Just as they selectively embrace only the Science that created their modern conveniences, while rejecting the Science that proves The Truth of Evolution; making them all intellectually dishonest, self-deluding hypocrites. All the BBA inflicted suffering on this planet is rooted in their 'denial' of The Truth of Evolution.

Only by embracing The Truth of Evolution may BBAs transcend the limitations of our Simian heritage, to achieve *enlightened self-awareness* derived Moral Justice and lasting Global Peace. Humane Sapients (HSs) embrace The Truth of Evolution as self-evident. And believe that if our fellow BBAs can't see that self-evident Truth in the mirror, they must suffer a severe intellectual deficit. If they don't even possess the simple rational judgment to correctly determine The Truth of *what they are,* and *where they came from,* and *how they got here;* a profound self-ignorance which deeply impacts IQ; how could any HS reasonably trust their judgment in anything else? The future HS society will never entrust *non-self-aware* BBAs with any 'authority', as all Humane Sapients' political and financial constructs will be founded upon the acceptance of The Truth of Evolution.

"Perspective is worth 80 IQ points." - Alan Kay

That enlightened self-awareness changes everything; and once realized cannot be unlearned; as a self-aware Humane Sapient can never again view this world as a non-self-aware Big Brained Ape. Once self-aware, they will forevermore judge the veracity of everything they see and hear through the *prism* of Evolution. Are large Ape brains capable of *'divination'*? Are even larger elephant and whale brains too? Do all Apes have *'souls'*? Does language make them *'eternal'*? Or just liars?

Every *'mystery'* in the Universe can be explained by Evolution; none are explained by false Origin Stories. There are no mysteries; only subjective ignorance and yet undiscovered objective Truths. Cast off all your Bronze Age ancestors' fear based 'faiths'; courageously seek the Scientific Truth.

"All Truth passes through three stages:
First, it is ridiculed. Second, it is violently opposed. Third, it is accepted as being self-evident."
-Arthur Schopenhauer-

NOTES

RELIGION

Teach your children that all 'religious faith' is rooted in delusional fear of the imagined. Fearing an imagined *'afterlife'*, Bronze Age BBAs constructed elaborate false mythologies to self-delude and comfort. The Greeks claimed some were 'Gods'; Roman and Egyptian 'leaders' claimed they were; while the modern 'faithful' claim they have an *'eternal personal relationship with God'*. Yet weekly they pay fellow lying Apes to scare them again, and once frightened then beg for reassurance that they aren't really just BBAs; whom, like their cousins in the forest will one day lie down and die, and return to the soil from whence they came. It's cowardly, irrational, mass grandiose delusion.

*'In all the Universe you alone are the unique eternal creation of the Divine Mind,
and all that you see around you was created solely for your use and enjoyment.'*

Lying to children is a form of abuse. Don't abuse your children; teach them *The One Proven Truth*. To persist in a 'belief' in spite of all Empirical Evidence to the contrary is a form of mental illness. To maintain that 'belief' when its purveyors rape and murder children is definitive proof of such.

The 'faithful' often cite the antiquity of their Origin Story as proof of its Truth; when 'The Method' proves the exact opposite is True, by constantly revising our *perspective* based on *new* evidence. Estimated to now double every decade; our species' Empirical Knowledge is like a living, organic, constantly growing 'body' of Truth. 'The Method' proves the *old* isn't 'sacred'; it's demonstrably in error, or at the very least, incomplete. Teaching children to structure their lives based on archaic 'Holy Books' imagined by Scientifically illiterate, cowardly superstitious primitives, is as absurd as teaching them to service their modern automobiles using a Model-T shop manual. The 'faithful' don't even know that no theme in their Bronze Age Harry Potter novel 'Holy Book' is even original; from the Garden of Eden, to The Flood, to their Messiah's 'virgin birth' and 'ascension'; everything they all 'believe' to be newly revealed is actually 'borrowed' from even earlier archaic Astrological traditions. Teach your children to be very wary of the BBA 'faithful'; for unlike Humane Sapients who honor The Truth; the 'faithful' have an emotional investment in their 'faith', they don't seek The Truth, they seek comfort. And as they lie to themselves, they will also lie to you, for BBAs never rise to the HS level of Intellectual Integrity, but rather HSs must always 'dumb down' to the BBAs' level instead. Decide when your children are very young, whether to teach them to structure their lives based on ignorance and fear and unethical *'absolution'* for their misdeeds; or on The Truth and courage and taking Personal Moral Responsibility for all their actions and treatment of all other life.

"We abuse the land because we regard it as a commodity belonging to us. When we see land as a community to which we belong, we may begin to use it with love and respect." - Aldo Leopold

An extreme danger of 'faith' is its proclivity to acclimate the 'faithful' to blindly ceding 'authority' to other BBAs. Opposite the 'The Method'; 'faith' fosters a heightened credulity and unwarranted trust in other BBAs' abilities and intentions; a potentially lethal miscalculation in many of life's situations. Always remember; no matter what their 'credentials'; they are just Big Brained Apes. 'Faith' fostered unwarranted trust was the Nazis' greatest ally, and enabled the pandemic of child raping Catholics. If there be a God who wanted 'fellowship' with our species, wouldn't all BBAs know it? Would an 'Omnipotent Being' need to send child raping Catholics and child murdering Muslims as Her emissaries to all the rest of us? Real 'faith' in such is mental illness manifested.

"Doubt is not a pleasant condition, but certainty is absurd." "Anyone who has the power to make you believe absurdities has the power to make you commit atrocities." "In the midst of all the doubts, which we have discussed for 4000 years in 4000 ways, the safest course is to do nothing against conscience. With this secret, we can enjoy life and have no fear of death."
-Voltaire-

NOTES

ATHEISM

'Faith' is the lazy minded weak default position that cowardly non-self-aware BBAs fall back to whenever confronted with the unknown. Finding The Scientific Method too much effort, they default to an imagined mystical worldview. Contrary to Truth honoring HSs who courageously admit that there are answers yet unknown, and maybe even unknowable in our Universe, which will require much more very hard work before The Truth may be conclusively confirmed, if ever.

The absence of evidence is not evidence of absence. The complete absence of any Empirical Evidence of the existence of 'God' is not proof that She does not exist, as one cannot definitively prove an absence, but only show that one has not yet found proof of a presence. Teach your HS children to never confuse their own subjective ignorance for objective proof of anything, much less The Truth. Ignorance is not evidence! Apply 'The Method'! When HS Thomas Edison was asked if he was discouraged by his many *'failures'* to successfully invent the incandescent light bulb he brilliantly replied: *"I have not 'failed'; I just found ten thousand ways that won't work!"*

Teach your HS children that in the complete absence of any evidence either way, agnosticism will remain the only ethical position of Intellectual Integrity for Humane Sapients, as all HSs practice epistemological modesty and don't presume to know what can't be known. While BBAs fill their children's heads with so many lies and nonsense that without their later adult reasoning applied through 'The Method', most will never realize The Truth. Humane Sapients teach their children *how* to think, rather than *what* to think; empowering them to discern The Truth for themselves. Enabling them to figure out at their own right time that 'Jesus' and 'Santa' are one and the same.

The BBA 'faithful' all assert that they have an *'eternal soul'*, yet amusingly they're unable to wrap their Ape brains around just a few thousand years of genetic isolation being able to create species' divergence through Natural Selection. They 'believe' in the micro-evolution of animal husbandry, yet deny the macro-evolution of species' divergence; not understanding that they're exactly the same thing extended over a much longer time span they ironically can't grasp. They 'believe' that hundreds of generations can turn wolves into chihuahuas, yet deny thousands can turn feathered lizards into birds? Limited by this ironic ability to conceive of *'eternity'*, but not of *Geologic time*, even the 'faithful' who falsely claim they accept the mutually exclusive Truth of Evolution versus *'Creationism'* usually mean from several thousand years ago Biblical mud hut primitives to modern BBAs. They don't actually accept that endless Empirical Evidence strongly suggests that we could have arrived at this present day by descending from the original 'Big Bang' solely through natural processes following the immutable laws of Physics, Chemistry and Biology without any need for *'supernatural intervention'* of any kind at any time. They 'believe' in fallacious Origin Story 'Holy Books' contradicted by all evidence, yet they deny endless Empirical Evidence based Evolution? Their delusional 'beliefs' in lies and nonsense don't advance our species' progress, they retard it; as to deny Evolution today is equivalent to denying forensic convictions with crime scene DNA.

Thanks to 'The Method', we know with certainty that none of the world's Bronze Age Harry Potter novel 'Holy Books' imagined anthropomorphic Deities exist. But for Apes who climbed down out of the trees just a few million years ago in a 14 billion year old Universe to assert absolute knowledge of God's non-existence in the total absence of any evidence is Simian banality; being as unprovable as the 'faithful's absolute claims of Her existence. The Universe is not limited to Apes' ability to understand it. We were oblivious to microbes prior to the 1673 Leeuwenhoek Microscope; *we didn't even know what questions to ask.* Atheism is just another time wasting 'faith'; while 'The Method' demands evidence. And any BBA who 'believes' 'The Method' and 'faith' can coexist understands neither. Get to work!

"The most costly of all follies is to believe passionately in the palpably not true.
It is the chief occupation of mankind."
-H. L. Mencken-

NOTES

DEATH

Death is the great equalizer. Every living thing dies. No exceptions.
Keep death as a constant advisor; your time here is preciously limited.

"Remembering that I'll be dead soon is the most important tool I've ever encountered to help me make the big choices in life." - Steven Jobs

Life is risk. BBAs cannot eliminate risk through legislation, they can only eliminate freedoms. Despite all BBAs' freedom limiting laws, always aimed at the exception to protect the less aware among us from themselves, Natural Selection will continue to cull them younger at higher rates. Every variable in Nature rides on a 'bell curve' and there is nothing BBAs can do to change that. There will always be the few percent of 'lunatic fringe' on either extreme that will always remain beyond our ability to help, and all of our attempts to help will only negatively impact the norm. BBAs will never dull the sharp culling edge of Natural Selection through Liberty limiting laws.

Where religion is delusional fear of a fantasy '*afterlife*'; Science is the search for Empirical Truth. Teach all children the Scientific Method proven Empirical Truths we now know of the immutable laws of Physics, Chemistry and Biology to empower them with the knowledge needed throughout their lives to make more judicious choices about their own safety and survival. Teach them the finality of death whenever it comes, not the fantasy of '*eternal awareness*'. No Apes live forever. None have '*souls*'. Big brain or small, their death is the end of their consciousness. And none are capable of '*divination*', and the sooner all BBAs realize that Truth, the less likely our extinction. Teach them to create their own joy filled 'Shangri La' *here and now*, for none awaits them later. They sell us later 'heaven' to enslave us here and now; the only here and now we'll ever know.

"Do not fear death so much but rather the inadequate life." - Bertolt Brecht

All consciousness is electrochemistry; the brain dreams the '*mind*', the '*mind*' dreams the '*soul*'. All death is brain death. Our brain is an electrochemical processor that simply ceases to function when either its current or chemistry is too severely altered. When our mortal organic brain dies, all its 'dreams' die with it, as dreams don't survive the dreamer, except through others still living. The fact that all '*near death*' experiences are very nearly identical proves that they are an organic, rather than a spiritual phenomenon. Oxygen deprivation is known to cause hallucinations, just as severe trauma can trigger the dream state of REM sleep, and the 'tunnel vision' also results from the way our retinas react to hypoxia. Obviously, our mortal organic brain has evolved a defense mechanism which creates pleasant, non-threatening hallucinations as its lights go out forever. And our obsession with our fellow BBAs' carcasses after death is disgusting. Displaying them and dragging them around in parades and wasting space burying them in prime ground. All too late. Whatever their legacy, it hasn't anything to do with the decomposing carrion in the casket. After a definitive forensic review every Big Brained Ape's carcass should be quickly cremated.

Teach your HS children to appreciate the here and now conscious, before they're forever gone. Teach your children this Truth; the greatest regret the dying feel is of the love they didn't express. Encourage your children to express their affections to all whom they love every chance they get. They may not get another opportunity, inevitable death may come swiftly and without warning.

"I cannot imagine a God who rewards and punishes the objects of his creation, whose purposes are molded after our own; a God in short who is but a reflection of human frailty. Neither do I believe that the individual survives the death of his body, although feeble souls harbor such thoughts through fear or ridiculous egotism's."
-Albert Einstein-

NOTES

ETHICS

"Respect for The Truth comes close to being the basis for all morality." - Frank Herbert

Would it ever be possible to construct a viable universal Code of Ethics based upon lies and nonsense? When the 'faithful' lie to their children and teach them; *'In all the Universe you alone are the unique eternal creation of the Divine Mind, and all that you see around you was created solely for your use and enjoyment';* how could they reasonably expect their children to behave ethically toward all their fellow Apes and every other species here on Earth? As all the BBA 'faiths' are based on imagined lies; it is no wonder none can get along with any other; as all their lies have no Truth based commonality. The 'Golden Rule'; *'treat others as you would have them treat you';* seems reserved only for members of their own 'faith'; while they often treat all others with an evil *'us'* and *'them'* bigotry; the prideful 'stupidity of ignorance'. BBAs claim they're of a 'faith', 'race' or 'nation'; but Evolution teaches that we're all children of the same planet. We must stop teaching our children lies and nonsense; as all BBA conflict here is rooted in their *learned* 'faith', 'race' or 'nation' based *'us'* and *'them'* bigotry.

"You can safely assume that you have created God in your own image when it turns out that God hates all the same people you do." - Anne Lamott

But if we teach all young children The Truth about their True Evolutionary commonality, they'll structure their lives based on an Ethical Code rooted in self-awareness derived humility, empathy, and compassion for all of their fellow evolved life; regardless of their ancestry. In a word; Dignity. If they need Bronze Age 'Holy Books' to teach them right from wrong, and how they should treat all other life; it's not 'faith' they're lacking, but an enlightened self-awareness derived empathetic Dignity. If we simply teach all young children their Evolutionary Truth, they've no need of Bronze Age myths.

<u>The Humane Sapient Universal Code of Ethics</u>
Always preserve another's Dignity, as you would have them preserve yours; never lie, cheat or steal, nor harm another's person or property; and never tolerate those behaviors in others.

And Evolution teaches them that all life deserves that same respect; as every species has *earned* the right to be here. HS society will punish the abuse of any life; requiring pet owner testing for worthiness before licensing them, instead of all their pets. And as sociopathic 'medical research' on Primates is torturing the equivalent intellect of a BBA child; it'll bear equal punishment to such. HSs will never permit the capture of any Big Brained species for 'entertainment'; as their confinements have always proven to be abysmal failures. Research now shows elephants and whales have complex vocabularies and emotional bonds to rival our own, we BBAs were just subjectively ignorant of them. All elephants know they're elephants; all whales know they're whales; and all the other Apes know they're Apes; we're the only Big Brained species deluded about our own objective Truth. Our Dignity derives from our Truthful self-awareness. One's self-awareness derived Moral Evolution can be judged by how they treat all other evolved life.

"Compassion for animals is intimately associated with goodness of character, and it may be confidently asserted that he who is cruel to animals cannot be a good man." - A. Schopenhauer

Personal Moral Responsibility for all of one's actions is the foundation of the HS Code of Ethics. While the BBAs commit atrocities in the name of their delusional 'faiths', only to be *'absolved'* of their immoral actions by their imagined deity; then claim we'd not know 'morality' if not for their 'Holy Books'; as they suffer undignified moral blindness from their lack of self-awareness. Unlike the non-self-aware BBAs, self-aware HSs don't rape or murder children; because it's undignified.

"Men never do evil so completely and cheerfully as when they do it from religious conviction."
-Blaise Pascal-

NOTES

SONS OF ABRAHAM

"A fact never went into partnership with a miracle. Truth scorns the assistance of wonders. A fact will fit every other fact in the Universe, and that is how you can tell if it is or is not a fact. A lie will not fit anything except another lie." - Robert G. Ingersoll

Before your children waste even the least of their preciously limited time here reading the false Origin Stories of the archaic Sons of Abraham (SOA), have them examine their current cultures. If they themselves derived no benefit from their own myths, how could your children reasonably expect to? How have their proven false Origin Stories served their creators' own cultures? Have they fostered everlasting Peace, prosperity and happiness; or incessant generational hostilities? Sometimes murdering those that didn't share their delusions, and sometimes being murdered by their delusional neighbors, ever since their imagined Origin Stories were first written; back when they were profoundly ignorant mud hut Apes, whose entire mythology was based on Bronze Age Astrology, while 'believing' that their flat world was the center of the Universe, *'created'* solely for them. Do any of the modern 'faithful' even question why the Pope wears a Pisces fish head mitre?

The prehistoric Archeological record contains evidence of continent wide trade between many contemporary cultures with little evidence of hostilities between those groups until after the SOA fabricated their divisive, delusive myths. Then, like a virulent Reason slaying high fever *infection;* BBAs have engaged in irrational hostilities everywhere it has spread around the planet ever since. If my children lived near any deluded SOA, I'd give them the same advice I'd give to anyone with irrational neighbors; *'Move!'* There is no 'Holy Land', there are no 'Holy Sites', they exist only in BBAs' delusions. It's all just dirt, go cultivate another patch, or they will give you no Peace ever. Whenever more SOA commit more atrocities, BBA Media always refer to them as *'extremists'* or *'radicals'*, but they're neither. They're simply *'true believers'* following their fallacious 'Holy Book' to the letter, and that is why one never hears of their clerics denouncing their mentally ill actions, because they would be denouncing themselves and the core body of their 'faithful' followers too. Their perniciously ignorant, prideful stupidity would be *tediously boring;* if it weren't homicidal.

We can no longer afford to allow those slow 'special needs' BBAs to dominate our species' global attention and agenda; we simply have too much *adult* work to do. SOA's fallacious, demonstrably destructive Bronze Age myths have brought all their own creators; and now all the rest of us; only endless hostilities and suffering. Thanks for sharing guys, but no thanks, you can keep *your* 'God'.

While Truth honoring HSs struggle to stay focused on our perilously uncertain road ahead, all those ever bickering 'faithful' in the back seat still dwell in our darkly ignorant Bronze Age past; without any knowing The Truth; their homicidal Simian rampages will likely hasten our species' extinction. Knowing *The One Proven Truth,* all HSs have then a Moral Duty to propagate that Peace inducing Empirical Knowledge by eradicating through education all the divisively irreconcilable, fallacious Bronze Age Origin Stories wherever they're still taught today. Who should set our species' agenda? Why have we allowed those with the least Intellectual Integrity among us to control our destiny? HSs must succeed in educating all those *myth infected* Apes, for lasting Peace will always elude us as long as there are BBA 'faithful' on this planet that fear The HS Truth more than their own deaths. Until all BBA myth infections are 'cured', every Ape will remain some other deluded Ape's *'infidel'.* All their *'Prophets'* and *'Messiahs'* are just dead Apes; and when the last of their 'faithful' die, their 'faiths', existing only in their deluded mortal Ape minds, will die with them; but we'll still be Apes. And if all Apes become extinct, all the world's 'faiths' will die with us; but The Truth will remain.

"We have enslaved the rest of animal creation and treated our distant cousins in fur and feathers so badly that beyond doubt if they were able to formulate a religion they would depict the Devil in human form."
-William Ralph Inge-

NOTES

SONS OF GREAT BRITAIN

"An invasion of armies can be resisted; an invasion of ideas cannot be resisted."- Victor Hugo

Historians universally condemn European Imperialism for its abuses; and even Gandhi, one of the most highly evolved to yet walk this planet, railed about the *'evils'* of British rule of India. Which shows the typically provincial nature of even the most enlightened BBAs' worldviews. Great Britain did exploit their conquests' natural resources; but what did they receive in return? It could be argued that Great Britain exported 'modern civilization' to the rest of the BBA world. Bringing their Rule of Law and judicial morality, modern state infrastructure and administration, roads and rail lines, institutions of higher learning and The Scientific Method. Thereby allowing all those lands formally occupied by BBA primitives, just two steps out of the mud, to leap many centuries ahead in their social, cultural and Scientific Evolution to their, and all of our, great and continuing benefit; as two hands working can do more than a thousand clasped in prayer; leaving a legacy of Indian colleges which today produce among the world's finest software engineers.

"Science has done more for the development of Western Civilization in one hundred years, than Christianity did in eighteen hundred years."- John Burroughs

Teach your HS children that it's no coincidence that the American Constitution, arguably the most enlightened structure of 'limited' government yet advanced by our species, was written exclusively by the Sons of Great Britain. And it's also no coincidence that Darwin's brilliant, world changing HS insight into our True Origins, was spawned by Great Britain's respect for, and support of, our species' unencumbered free intellectual inquiry; which is the beginning of all other Liberty; as no BBA can ever be truly free, until they know The HS Truth; acquired by applying 'The Method'. How different would our world be without just those two Great Britain inspired advances alone? Their Truths will last; others' lies won't. HSs will honor the American Constitution and Darwin's insight forever, while all the world's Bronze Age 'Holy Books' will eventually fade into obscurity.

"The future has already arrived. It's just not evenly distributed yet."- William Gibson

Courage and Trust; the underlying character principles of The American Constitution and The Scientific Method. Great Britain had the Courage to Trust themselves and their fellow citizens with unprecedented freedom of choice, and freedom of action, and freedom to inquire; as they perceived was best; in the pursuit of unprecedented knowledge and enlightened self-awareness. Teach your children that Great Britain's spark of Moral Courage permanently illuminated our then dark world to the continuing benefit of all current and future BBAs everywhere on Earth. Among all the world's family of nations, Great Britain shall always deserve an esteemed position, as never again are we likely to witness their unique contribution to our species' enlightened self-awareness and resulting Moral Evolution. All the world's nations must now follow their courageous example to attain their own Empirical Evidence based Humane Sapient Truth *perspective* to truly progress.

Ask your HS children how different would our current world be if the Spanish with their Catholic Inquisition had colonized North America as well; instead of the British with their Magna Carta? Ask them how Catholic South and Central America's myth infections have retarded their progress? Ask them to imagine how much further advanced our species would be had the Sons of Abraham never written their Reason slaying Bronze Age Harry Potter novel 'Holy Books' in the first place? For Great Britain's Son Darwin so loved the world that in an Age of universal deceit, he had the Courage to Trust himself to fearlessly seek The Truth to advance our species' Moral Evolution.

"There is no Truth existing which I fear, or would wish unknown to the world."
-Thomas Jefferson-

NOTES

FOUNDING FATHERS

Many modern Christians immorally claim that we should incorporate their lies and nonsense based 'Creationism' and 'Intelligent Design' arguments into all our public school curriculums because our Founders shared their 'beliefs'. Although many of our Constitutional Founders were indeed Deists, they were not, in fact Christians. And as their philosophies were developed a full century before Darwin's brilliant insight was widely peer reviewed among all his Scientific contemporaries; by reading our Founder's own words; it is obvious they would have also *'held'* Darwin's Empirical Evidence based Evolutionary Truth *'to be self-evident'*, once they had reviewed it themselves.

"I have found Christian dogma unintelligible. Early in life I absented myself from Christian assemblies." "I cannot conceive otherwise than that He, the infinite Father, expects or requires no worship or praise from us, but that He is even infinitely above it." "The way to see by faith is to shut the eye of Reason." "Revealed religion has no weight with me."
-Benjamin Franklin-

"As to the Christian system of faith, it appears to me as a species of atheism; a sort of religious denial of God. It professes to believe in a man rather than in God. It is a compound made up chiefly of man-ism with but little deism, and is as near to atheism as twilight is to darkness."
-Thomas Paine-

"Experience witnesseth that ecclesiastical establishments, instead of maintaining the purity and efficacy of religion, have had a contrary operation. During almost fifteen centuries has the legal establishment of Christianity been on trial. What have been its fruits? More or less in all places, pride and indolence in the clergy, ignorance and servility in the laity, in both, superstition, bigotry and persecution."
-James Madison-

"The clergy believe that any portion of power confided to me, will be exerted in opposition to their schemes. And they believe rightly; for I have sworn upon the alter of god eternal hostility against every form of tyranny over the mind of man." "Shake off all the fears of servile prejudices, under which weak minds are servilely crouched. Fix Reason firmly in her seat, and call on her tribunal for every fact, every opinion. Question with boldness even the existence of God, because, if there be one, he must more approve of the homage of Reason than that of blindfolded fear." "And the day will come, when the mystical generation of Jesus, by the Supreme Being as His Father, in the womb of a virgin, will be classed with the fable of the generation of Minerva, in the brain of Jupiter."
-Thomas Jefferson-

"The question before the human race is, whether the God of Nature shall govern the world by his own laws, or whether priests and kings shall rule it by fictitious miracles." "...this would be the best of all possible worlds, if there was no religion in it."
-John Adams-

Our Founders declared there'd be *'no religious test to qualify'* for any public office or trust; implying that 'authority' candidates should qualify for public office or trust by their Rationality. Spend Sunday mornings studying the American Constitution, the Federalist Papers, and everything else ever written by all the HS genii above with your children, not Bronze Age lies and nonsense; as Liberty is never more than one generation away from extinction. Or maybe our species is not yet worthy of Liberty and Self-Determination? Perhaps BBAs need yet another few centuries with a boot on their throats?

"Liberty cannot be preserved without a general knowledge among the people."
-John Adams-

NOTES

GOVERNMENT

"It is the first responsibility of every citizen to question authority." - Benjamin Franklin

Humane Sapients will require our elected servants to apply The Scientific Method to prove their assertions, and to swear to support and defend The HS Code of Ethics along with our Constitution; instead of the current BBA practice of electing them based on their chosen false Origin Story 'faith'. And HSs will institute Scientific peer reviews to nullify any existing laws that remain on the books just because they benefit our elected servants' friends, as unlike HSs, many BBAs value profit above their own integrity. To attain The Age of HS Enlightenment, it's crucial we only elect Scientifically literate HSs who honor Empirical Evidence, the Truth of Evolution, and the HS Code of Ethics it supports, to create all policies on our behalf. While the BBAs' current foreign policy structures our alliances with other nations based on no reason more sane than that they 'believe' the same Bronze Age Harry Potter novel 'Holy Book' is the *'inerrant word of God'* as most of the U.S. electorate; not on their true strategic value to our nation's Liberty and Security, and often acutely contrary to both. The 'Holy Land' exists only in the minds of deluded BBAs, yet America expends billions annually 'defending' it; all the while alienating surrounding neighbor populations to our extreme detriment. An insane aggravated menacing strategy considering our current dependency on their petroleum. Remove all 'faith' from that equation and U.S. policy towards that insignificant sliver of worthless desert is equivalent to Russia expending billions annually 'defending' the Navaho Reservation in Arizona. Just how would all Americans living in Arizona, Utah, Colorado and New Mexico react? To compensate for that insane policy, we must then also pay all their neighbors billions annually in 'Peace extortion'; thereby funding *an endless immoral cycle* of deluded BBA 'faith' based conflict.

The American Constitution could, and should be, The HS Truth 'Light of The World'. Our only exports should be food and The HS Truth. Feed the hungry and educate the ignorant. We should continue the enlightening work of Great Britain, not only for our own increased security, but to empower all BBAs with the self-awareness derived Dignity needed to increase their own security by ability to self-govern. We had Britain's Courage to Trust Germany and Japan in the Marshall Plan and they became our strongest allies and trading partners. We must do the same for every nation because The Truth knows no country. Only then may we achieve True Justice and lasting Global Peace. America must abandon all of our delusive 'faiths' and *lead by Rational example.*

"May the Declaration of Independence be to the world what I believe will be (to some parts sooner, to others later, but finally to all), the signal of arousing man to burst the chains under which monkish ignorance and superstition has persuaded them to bind themselves, and assume the blessings of security and self-government." - Thomas Jefferson- 1826- last known writing.

Instead, we betray *the promise* of our Constitution by supporting regimes that repress the will of their own people for self-government. Our cowardly BBA servants' fallacious 'faiths' delude them into repeatedly backing the wrong horse, and then they're surprised when those repressed espouse Jefferson's life motto; *"Rebellion to tyrants is obedience to God!";* as yet another war erupts. It's mental illness manifested. 'Progressives' claim our immigration system is *'broken'*, yet we process more immigrants on the road to full legal citizenship every year than all other nations combined. Those immigrants honor our HS Constitution more than our own cowardly elected servants, who deludedly presume to dismantle the very Civil Liberties those immigrants seek, for the illusion of 'safety'. Their actions prove their simple BBA intellects completely misunderstand our brilliant HS Founders' intentions, because they lack their Courage to Trust. Only the cowardly 'faithful' imagine government can 'protect' them. HSs honor Liberty, and possess the Courage to Trust others with it.

"A man may die, nations may rise and fall, but an idea lives on."
-John F. Kennedy-

NOTES

DNA

"What Science promises, and has already supplied in part, is the following. There is a real creation story of humanity, and one only, and it is not a myth. It is being worked out and tested, and enriched and strengthened, step by step." - Edward O. Wilson

The fossil record shows there were once many types of BBAs living concurrently on this planet. Rather than a linear descent, think of a many branching tree. Including us, at least four branches of BBAs lived here as recently as just 30,000 years ago; three until 24,000; and two until 12,000; a mere blink of an eye in Geologic time. Two of those 'cousins'; had survived for much longer than our own distinct branch has so far to date, but for many reasons including climate change, all of the other branches of successful BBAs eventually became extinct except us. We were not unique, we just better adapted to our changing environments, overwhelming them with more offspring. By tracing female mtDNA, we know that every current BBA on this planet is descended from a single African family living more than 150,000 years ago. And by tracing a marker on the male Y chromosome we know that the earliest their descendents likely migrated out of Africa was less than 75,000 years ago. Likely in small extended family groups eastward into the huge southern Eurasian savanna where we then bloomed into millions as we spread out across the entire planet. Teach your HS children this profound Truth; all BBAs on Earth are one single extended family. Take them to the 'Hall of Man' at your local Natural History museum, and let them see and touch the fossil record for themselves. Then take them to the 'Great Ape House' at your nearest Zoo and show them their magnificent relatives. Teach them the profound Truth that BBAs share a quarter of their Genome with yeast, indicating a shared single celled common ancestor. Take them all as young children, before they're exposed to the many lies of the 'faithful', as all children intuitively recognize The Truth when they see it, and that knowledge will empower them for the rest of their lives to defeat the many evils of fear based imagined Origin Stories and other *'us'* and *'them'* lies.

"It is easier to build strong children than to repair broken men." - Frederick Douglass

That same intuitive recognition of the hypocrisy and injustice of the *legal fiction* financial world the BBAs who came before them have constructed is also the source of their teenage rebellion. So teach all children The HS Truth and their Moral Duty to correct those unethical constructions from the time they're young adolescents, which will eliminate much of their bitterness throughout the difficult teen years of their increasing awareness of the BBA constructed world's immorality. And if BBAs' flagrantly unethical and unjust constructs don't still make you angry, you are just no longer paying attention the way they do, and you once did. Teach them they can achieve whatever they choose to achieve; and that nothing, and no one else is going to correct them if we HSs don't.

"When you grow up, you tend to get told that the world is the way it is, and your life is just to live your life inside the world; try not to bash into the walls too much, try to have a nice family life, have fun, save a little money. Life can be much broader once you discover one simple fact, and that is; everything around you, that you call 'life', was made up by people that were no smarter than you. The minute that you understand that... that you can change it... you can mold it... that's maybe the most important thing." - Steven Jobs

Teach young children this profound Truth; there is no *'us'* and *'them'*, there is only *'we'*. Only BBAs' learned 'faith', 'race' or 'nation' based *'us'* and *'them'* bigotry separate us; and they have a Moral Duty to correct the errors and unjust constructs of all the self-ignorant Apes that came here before them. Teach them; *True self-knowledge* is the beginning of all Wisdom; *'we'* are all one family.

"The chief cause of human error is to be found in prejudices picked up in childhood."
-Rene Descartes-

NOTES

RACE

"Condemnation without investigation is the height of ignorance." - Albert Einstein

Teach all young children that there is simply no genetic justification for 'racial' classifications. The genetic profile of two Africans from the same region can be more diverse from each other on average, than either is from a randomly chosen European or Asian. This is partly explained by the fact that all BBAs originated in Africa; making all African populations older, and therefore more diverse. But it also shows how closely related we all truly are to each other. We are all one family. All the superficial external 'differences' between BBA populations are simply their adaptations to their ancestral environment's unique pressures on their population. Once your children understand that HS Truth, they will marvel at their fellow children's ancestors' adaptations to their historical geography, rather than fear them because they are 'different'. Hate is rooted in fear, fear is rooted in ignorance. Teach every child while young to always; Listen Less and Think More. (LLTM)

"Knowledge is the antidote to fear." - Ralph Waldo Emerson

Skin color results from the amount of *melanin* in the skin; a beautifully functional protective adaptation to deadly Ultra-Violet Solar radiation which can mutate chromosomes and destroy folic acid necessary for proper fetal development. As additional DNA protection, every female BBA is born with all the 'eggs' she will ever bring to maturity already inside her undeveloped ovaries, transferred directly from mother to daughter, while in the safety of her mother's womb. The closer one's ancestors lived to the equator where the UV radiation is most intense, the more melanin they evolved in their eyes, hair and skin; excepting the unradiated bottom of their feet. Insulating curly hair and a flatter nose are also protective by reducing UV exposure from directly overhead. *'Form follows function'*. UV radiation also causes cataracts, so always wear sunglasses. As others' ancestors migrated toward the poles where the UV radiation is most scarce; a reduction in melanin content was 'selected for' to facilitate the absorption of as much of the limited sunlight as possible, to create precious Vitamin D necessary for proper bone development. And their noses became larger to filter and preheat the frigid air as it was inhaled into tender internal membranes. BBA populations from intermediate temperate regions share characteristics of both adaptations, making the ancestral BBA color spectrum run from blue black to pale white and every shade in between. Clearly BBAs can only thrive in a narrow range of UV radiation absorption and adapt accordingly. As BBA populations become less geographically isolated, this Truth becomes less obvious, but assigning different classifications to populations based on melanin content is absurd.

'Race' was just another nefarious fabrication used to divide BBAs into *'us'* and *'them'* to justify our Simian aggressions. It is a classification unsupported by genetics, and if the ignorant persist in this fallacy today they should more appropriately be labeled as 'melanists', rather than 'racists'. And for any BBA to define themselves by their own melanin content, a trait over which they have no control, nor is relevant to either IQ or character, is a woeful testament to their self-ignorance. As our diverse BBA populations continue to intermix, through hybrid vigor our single family will become more heterosistically intelligent, attractive, athletic and creative, as the most talented and beautiful among us are often the offspring of parents of widely diverse geographical ancestries. In time, all our globally mobile family will be one medium amber skin hue with thick wavy hair. If melanin content is the absurd singular criteria for 'affirmative action', then by law shouldn't every person of Scandinavian descent who has a dark 'tan' qualify as a 'protected class' too? Why not just teach all young Big Brained Ape children The Humane Sapient Truth instead?

"We shall not cease from exploration, and the end of all our exploring
will be to arrive where we started, and know the place for the first time."
-T. S. Eliot-

NOTES

WAR

"Be wary of the man who urges an action in which he himself incurs no risk." - Lucius Seneca

Teach every young child you meet; war is murder; war is failure; war is a lie; war doesn't work. Delusional BBAs claim war represents our greatest attributes of *'honor'* and *'selfless sacrifice'*. Self-aware HSs know war represents the very worst attributes of our primitive Simian heritage. There is no honor in killing or dying, only wasted potential. And no fallacy is made Truth, just because some myth infected Apes will kill and die for it. If war worked, there'd be no more war. Can any BBA please show us where in our Constitution it mandates that we're supposed to go into other nations and covertly meddle in their politics, and then murder their citizens if they resist our directives? The initiation of force is immoral. Although less than five percent of our total global population, America's 'military' spending now far exceeds all other nations on Earth combined.

"Over grown military establishments are under any form of government inauspicious to Liberty; and are to be regarded as particularly hostile to Republican Liberty." - George Washington

BBAs lie to our children, dishonorably sending them to murder to achieve their own rapacious goals. How's that been working out for all the world's children so far? Teach all young children that Truth so they may defeat the BBAs' *'us'* and *'them'* lies, while working towards the lasting Peace that will grow from The HS Truth's global acceptance. All HSs must then lead by Rational example. The solution to conflict is to feed the hungry and educate the ignorant. But ever since America's righteous self-defense in WWII, we've seen our Military murder 'faithful' melanin rich *'them'* in their own lands all over our planet. We've become what we defeated. It never achieves lasting Peace; but only foments the exact opposite; by seeding the next generation of armed combatant *'them'* throughout their entire region. How many times must America infuse a region with arms; only to have them later used against *'us'* before we learn that none ever benefit from conflict, *except the financiers,* whose sole 'interest' is in owning and trading *the debt* run up on both 'sides'; their profits funded with innocent blood.

"There is no flag large enough to cover the shame of killing innocent people for a purpose which is unattainable." - Howard Zinn

We spend almost $1 trillion each year murdering others in their own lands. How did we get here? What if we just spent billions feeding all the hungry instead? Even young children know you can attract more friends with honey than vinegar. Who's country is this? What a tragic waste that our limited national resources are devoted to the development of mechanized murdering technology, when there's so much crucial work to be done. BBA children are starving; pandemics are raging; the environment is degrading; all of which could be solved by Science if they had our focus and funding, instead of the BBAs' machines of death and destruction. It is mental illness manifested. Thanks to BBAs' *'us'* and *'them'* delusive 'faiths', we are currently an extraordinarily sick species wasting this brief, extraordinary singular life experience, on an extraordinarily beautiful planet. Genetic Science has proven we diverged from the other Pan Apes more than five million years ago. Yet the best we can do is murder each other over inane 'faith', 'race' or 'nation' based *'us'* and *'them'* bigotries? All the world's children are watching! They'll pursue whatever we teach them by example. What might they *achieve together* without this constant immoral distraction and funding of conflict? How many now accept War as a natural state of being for our species? We must teach them all it isn't; by teaching all young BBA children on Earth the Dignity and Peace inducing Human Sapient Truth; as BBAs cannot murder our way to Reason, nor towards the advancement of our Moral Evolution.

*"Non-violence leads to the highest Ethics, which is the goal of all Evolution.
Until we stop harming all other living beings, we are still savages."*
-Thomas A. Edison-

NOTES

THE COLD WAR

America lost fewer than half a million citizens in all of World War II. Russia alone lost more than twenty million, over a third of all other 'combatant' nations combined. Many millions of European Russians were killed, it was the Asiatic Russians who finally stopped, and then drove the Nazis back across their vast lands into Germany. The equivalent of Japan attacking our west coast and driving in all the way to the Rocky Mountains before finally being driven back into the Pacific. Is it any wonder the Russians sought 'protective' land holdings in Europe to secure their western borders? Didn't we do the very same thing ourselves with Okinawa and the other Pacific islands? A full quarter of all 'Southern' men were killed in the American Civil War, and even after more than eight generations later, all 'Southerners' are still 'tender' about that conflict even today. Our judgment of others' motives and actions must be formed from *their perspective.* Imagine Russia's national paranoia after such a devastating experience. No wonder Khrushchev pounded the U.N. podium with his shoe. The Russians were all terrified of another conflict. Could you blame them? Just imagine what *'they'* may have thought about *'us'?* And how did America respond to their fears? Instead of reassuring them that we would never attack them and only wanted to live in harmony; with normalized trade and constant dialog; we threatened them with a massive military build up and surrounded them with 'intelligence' gathering nuclear armed bases and spy plane overflights. Then BBAs were surprised when *'they'* tried to do the very same thing to *'us'* in Cuba? HSs weren't. This is the proximate extinction legacy that BBA Simian military thinking has left to our children. While HSs know that all forms of tyranny; Theocracy to Communism; will only disappear from our planet through education, as Liberty happens in the mind first. All intellectual bondage will be unsustainable once the coming HS Age of universal enlightened self-awareness is achieved.

"It is Truth that liberates, not your effort to be free." - Jiddu Krishnamurti

All Empirical Evidence suggests that this singular life is the only time that any of us will be conscious. Unaware of *what they are,* and *where they came from,* and *how they got here;* our *deceived and abused* Military; all taught to obey the same lies, cloistered away from any rational Truth honoring peer review dissent, are oblivious of the path to advance our species' Moral Evolution. They waste this singular life murdering their fellow Apes, yet 'believing' they're murdering *'them'* for *'God and Country'* in our name. Is that not mental illness? They don't represent HSs. Are they murdering *'them'* for you? And they don't represent our country; they're only enforcing their financiers' profits. Is Vietnam any different now after a million of *'them'* and 60,792 of *'us'* were murdered there? 'PTSD' is nothing more than their shattering disillusionment; the shocking realization that they've been betrayed with lies for their entire lives. Until our HS Scientists control our BBA Military, all our planet's children will remain perilously close to our self-induced extinction; as Peace will only spread as enlightened self-awareness derived Dignity spreads.

"We can easily forgive a child who is afraid of the dark. The real tragedy of life is when men are afraid of the light." - Jo Petty

Imagine what we and the Russians could have *achieved together* if we'd focused all of our collective energies and Military budgets on Space Colonization over the many decades since WWII, instead of NASA's current budget of a tiny fraction of the BBAs' *immense wasted military spending;* which now alone accounts for a full quarter of America's total Federal spending. The wealthiest nation on Earth has now run up a $17 trillion *debt* by its immoral force initiating, myth infected BBA politicians.

"Every gun that is made, every warship launched, every rocket fired, signifies in the final sense a theft from those who hunger and are not fed, those who are cold and are not clothed. This world in arms is not spending money alone. It is spending the sweat of its laborers, the genius of its Scientists, and the houses of its children."
-Dwight D. Eisenhower-

NOTES

SPACE COLONIZATION

*"That which cannot be compassed by reason, wisdom and discretion,
can never be attained by force."*- Michel Eyquem de Montaigne

Teach every HS child; there are only four endeavors worthy of HSs at this stage in our Evolution. Feeding the Hungry; Educating the Ignorant; Preserving the Environment; and Colonizing Space. Everything else is Simian banality; *Evolutionary Stagnation.* Who's in charge here? We can only lead by Rational example, never through force. America's lost its moral compass and leadership.

BBAs evolved as a direct result of a mass extinction event sixty six million years ago when an object from space slammed into the Yucatan peninsula killing most of the other species on our planet by its long term, Sun blocking atmospheric debris. Giving our warm blooded tree shrew sized ancestors the opportunity to exploit all the niches left by all those other species' extinctions. The Earth is constantly bombarded by objects of all kinds from space and the fossil record shows that in the almost four billion year history of life on this planet there have been many similar mass extinction events caused by those objects occurring at fairly regular intervals. We know then with certainty that our planet will suffer many more such collisions, we just don't know exactly when. If our species is to survive the next mass extinction collision event, we must develop a 'lifeboat' of BBAs away from our Earth 'mothership', by establishing permanent colonies on other planets. The dinosaurs 'reigned' on this planet for 160 million years yet were powerless to prevent their own unexpected extinction from space. We have only survived for several million years to date; but with proper Empirical Evidence based planning and action, we are not equally powerless to prevent our own similar extinction. We must then make plans viewed through the prism of Geologic time. Our best hope lies in the colonization of first Earth's Moon, and then the terraforming and colonization of Mars and its Moons. If we HSs don't establish 100, 300, 500, 1000, 3000 and 5000 year plans of Rational action to ensure our own thriving survival, nothing, and no one else will do it for us. No HS endeavor will be more crucial to our species' long-term survival, and nothing will better serve to challenge the best of all our species' talents and imagination, than to terraform and then colonize Mars and its Moons. But while clueless BBAs continue to squander our finite 200 year *one time free pass* on cheap oil on their insane Simian aggressions, HSs know that we have just an instant in Geologic time before it's gone forever later this century, and we must switch our species' focus to Space Colonization now, before it's gone, or we may never get off this rock.

*"Now it is time to take longer strides. Time for a great new American enterprise.
Time for this nation to take a clearly leading role in space achievement; which
in many ways may hold the key to our future on Earth."*- John F. Kennedy

Just a fraction of BBAs' *immense wasted military spending* would be more than enough to fund this endeavor, as well as to feed and house and educate every hungry or homeless child on Earth. Until HSs make that happen, the Moon will remain a nightly HS reminder of our BBA stagnation. *'Thinking outside the box'* has become the box. Every BBA child should spend an hour every day *'Thinking outside the planet'* instead. Or should they focus backwards into our darkly ignorant past and continue our current plans of delusive Origin Story driven homicidal Simian rampages all the while 'praying' for Armageddon's Bronze Age fantasy *'afterlife'* until they exterminate themselves? Now there's BBA genius for you! And they think *'we're so sophisticated'* don't they? Stupid Apes. Imagine what they could achieve if every myth infected BBA spent an hour a day contemplating The Truth toward advancing our Evolution, instead of that same hour 'praying' for its violent end? If not for our fantasy *'afterlife'* obsessed religious retardation we could have been on Mars already.

"Nothing in the world is more dangerous than sincere ignorance and conscientious stupidity."
-Martin Luther King, Jr-

NOTES

HUNGER

"If the misery of the poor be caused not by the laws of Nature, but by our institutions, great is our sin." - Charles Darwin

Many thousands of BBA children starve to death on this planet *every single day*. While fat Chimps in suits force millions of meals worth of soft commodities to go to waste or to animal forage every single month to 'fix prices' in their immoral BBA constructed legal fiction economy. Even at over 7 billion, there's not yet a global shortage of food; only a global shortage of goodwill towards our own siblings.

"More money flows through the private capital markets in a day, than through all the world's governments in a year." - John Doerr

How can BBAs live with those Truths and think themselves as a moral species? Why isn't eradicating child hunger the BBAs' collective top priority? Who *profits* by their hideous suffering? The next time you're gorging yourself at your replete table, commit to going without any food for 24 hours once per week until we HSs end their grotesque agony. Where's the 'faithful's *supernatural intervention'* now? Get up off all your 'praying' BBA knees; nothing and no one else is going to do it if we HSs don't. It isn't *'God's will'*; it's *your* will. It's estimated we could end all child hunger on Earth with less than one percent of just the 'wealthy' nation's government budgets; yet we waste many times that every year on inane Ape distractions. We pay 'adult' BBAs many millions each year to 'play' simple games of children, all the while others' children are starving to death? After millions of years of Hominin Evolution is that really the best BBAs can do? Throw a ball through a hoop? Hit a ball with a stick? What would our archaic ancestors say if they could see our present stagnation? *'We barely survived starvation in caves, daily fighting off megafauna predators, so you could 'chase balls'? While your Scientific Community, who are your best hope for eradicating hunger, toil in unfunded obscurity?'* The only historical value of athletic competition was in educating the ignorant that we are all one single family despite our melanin diversity; achieved without the need of any 'professional' sports.

"Athletic sports, save in the case of the young, are designed for idiots." - George Nathan

Instead of BBA society's immoral funding of childish games, the future HS society will reward every advanced Scientific and Teaching degree by canceling their entire student debt upon their successful graduation; and a lifetime guaranteed high annual chosen field salary; to reflect their true value to our thriving survival. Our children will pursue whatever Evolutionary contribution we encourage of them by example. Science stars or sports stars? HS progress or BBA stagnation?

"Education is teaching our children to desire the right things." - Plato

A BBA graduates from college imagining they know much of what there is to know about their chosen field, an absurdity shared with the other self-ignorant 'faithful' BBAs within their discipline. While an HS graduates with a humbling awareness of just how much more there is yet still to learn. As Evolution teaches humility and empathy with all other evolved life; while all religions teach self-aggrandizing delusion. The HS Truth forever changes one's *perspective*; we have much work to do!

"We don't know a millionth of one percent about anything." - Thomas A. Edison

Teach every young child The Truth about BBAs' immoral speculative 'price fixing' for profit on food and to share our planet's finite resources with everyone in our melanin diverse single global family.

"To a man with an empty stomach; food is God."
-Mohandas Gandhi-

NOTES

EDUCATION

"It is far better to grasps the Universe as it really is than to persist in delusion, however satisfying and reassuring." - Carl Sagan

No endeavor is more vital to finally achieving lasting Global Peace and Justice and True Liberty than empowering all young BBAs with The HS Truth and The HS Code of Ethics it supports, as to treat one another with enlightened self-awareness derived empathetic Dignity is the only way to attain them. The conversion of every BBA child into a self-aware, humbly compassionate and empathetic HS adult will finally achieve what no Bronze Age BBA myth infections ever could; the permanent dignified end to all war, fear, hatred, hunger, bigotry and suffering on this planet.

"What we achieve inwardly will change outer reality." - Plutarch

Every HS adult has a Moral Duty to teach all BBA children they meet The HS Truth; teaching them *how* to think, rather than *what* to think; applying 'The Method' to discern The Truth for themselves.

"Education is the kindling of a flame, not the filling of a vessel." - Socrates

Give every child a microscope and telescope and encourage them to explore the Cosmos beyond all their five evolved senses. From the high alpine to the coral reefs, explore every environmental niche together; immersing them in the Natural World at every opportunity; for the enlightened self-aware HS Truth is there to instructively witness, they have but to let it in. Overcome the cowardly anti-Truth 'faithful' lobby active in many school districts throughout this country. Do all you can to defeat their lies and nonsense and improve the quality of the Science curriculum in all our schools. And do all you can to increase Teacher compensation so that we may attract the best HS educators for all our children, as a war between modern Scientific Truth and ancient Bronze Age mythology rages for every vulnerable young mind. Of all pursuits, Teaching The HS Truth is the most noble.

"A Teacher affects eternity; they can never tell where their influence stops." - Henry B. Adams

Bring all children you meet into the light of the self-aware HS Truth, help them finally banish all false mythologies to our darkly ignorant past. As ignorant myth infected Big Brained Apes, their extinction is inevitable. As enlightened self-aware Humane Sapiens, their potential is unlimited. Since Darwin first discovered The HS Truth, all BBAs have 'believed' he was wrong. What every BBA child must then ask themselves; by spending an hour every day contemplating the Natural World through the prism of Evolution is; *'Just how ignorant must one be to 'believe' that Bronze Age primitives understood our Universe better than we do today?'* Extinction is the norm in the fossil record; over 99% of all the species that ever lived on this planet are now extinct. And if we stay on this rock, ours too is certain; but we can do much to delay that inevitability. And the best way to ensure that is to teach every young child *The One Proven Truth,* instead of all the 'faithful's many Bronze Age lies and nonsense. We delude ourselves that we have 'made it' as a species, that we've 'conquered' the Natural World, and so spend much of our time on distraction; but The Truth is that our foothold here will always remain precariously fragile. Who could have stepped in the footprint of a dinosaur; in the middle of their 160 million year reign on this planet; and not have believed that *they* were the 'Apex of Evolution', and would continue to reign forever? Education is worthy of our greatest respect and passion, as Big Brained Apes must pass along our species' collective Empirical Knowledge derived HS Truth and the Dignity and Peace inducing HS Code of Ethics it supports to the greatest percentage of our children; their very survival depends on it.

"The great threat to the young and pure in heart is not what they read, but what they don't read."
-Heywood C. Broun-

NOTES

MOTHER EARTH

"The more clearly we can focus our attention on the wonders and realities of the Universe about us, the less taste we shall have for destruction." - Rachel Carson

The Earth doesn't need *'saving'*; it'll continue to circle our Sun until it 'burns out' and engulfs Earth in another five billion years. It is only our own survival, that we may, or may not *'save'*. Evidence suggests Mars once had an atmosphere and flowing water, but now is devoid of both. If BBAs alter our planet's atmosphere too dramatically, the future Earth may share Mars' fate. Whatever we do, we do to ourselves and to our fellow species; not to this rock. A great exercise for your child's daily hour of Natural World contemplation is to imagine space travelers arriving at the Earth for the first time. Would *they* have any doubt that 'Humans' are BBAs? Would your children ask them how they'd survived their species' adolescence to achieve interstellar flight? How would your HS children answer those alien's questions about BBAs' endless warfare, and starving children, and this generation's reckless squandering of the finite cheap oil while severely altering their own grandchildren's planetary atmosphere for their own selfish short term profit? All BBAs are blind to the damage because they don't see it through the prism of Geologic time. So a battle now rages between the HSs who possess *Geologic vision* and the BBAs who don't. Three generations ago one could get pure water directly from any stream; and CO_2 devouring primordial forests covered much of the planet; and global population was under three billion. If we stay on our present course, many of our great grandchildren will starve, as the greatest threat from the 35 billion tons of CO_2 that BBAs add to Earth's atmosphere annually may be the alarmingly quick lethal acidification of Earth's oceans. If we kill our planet's oceanic food chain, this rock will survive, but we're done; game over. What would those aliens ask us then?

Global weather is driven by only three variables; temperature, humidity and pressure gradients. Earth's atmosphere holds over six times more fresh water than all our planet's rivers combined. As our planet continues to warm later this century, the additional atmospheric moisture from the increased Solar evaporation of the oceans will continue to increase the severity and frequency of all 'storms'. Severe hurricanes, tornadoes, blizzards, droughts and fires will only get worse. And unless we preemptively relocate our metro populations out of 'tornado alley', and ten miles inland, returning all of our coastlines to undeveloped public beaches, the carnage will become extreme. We will never solve these problems as long as we leave control of policy to myth infected BBA politicians, who lack Geologic vision and aren't required to prove their assertions like Scientists. James Watt, the Interior Secretary under President Reagan, based U.S. Environmental Policy on his 'faith' that because *'Jesus is coming back'* in *'this generation'* and would *'create a whole new Heaven and Earth'*, we could ravage our forests and streams and oceans with impunity for selfish short term profits. This is the mentally ill, environmentally destructive legacy that deluded 'faithful' BBA thinking has left to all our children. Just for cheap electricity, that same myopic BBA insanity destroyed in just one generation the Pacific Northwest salmon runs; which took millions of years to evolve into one of the most reliably abundant high quality nutrient sources on our entire planet.

Teach your children by example; encourage conservation and recycling of every resource to them. Saving over 90% of the fresh water and energy required to make it from scratch; recycling just a single aluminum can or glass bottle saves enough energy to run a computer for more than an hour. Vote only for HSs who affirm the Empirical Evidence proving BBAs' environmental degradation, and commit to adopting policies that restore rather than further degrade our grandchildren's planet. Never vote for immorally avaricious, environmentally destructive BBAs with imaginary friends.

"The oceans are the planet's last great living wilderness; man's only remaining frontier on Earth, and perhaps his last chance to prove himself a Rational species."
-John L. Culliney-

NOTES

WATER WARS

Without a visionary American commitment to aggressive R&D funding of 'alternative energy' to eliminate our dependency on petroleum, the Pentagon; concerned for the long term security of our sovereignty; have a strategic 100 year plan in place to maintain our military 'presence' in the oil rich Middle East nations until they're pumped dry later this century. And as our myth infected BBA politicians lack the Geologic vision to address other threats to our sovereignty until they're already upon us, soon our deceived and abused Military will again be pressed upon to provide similar protections for an infinitely more precious rapidly diminishing finite natural resource.

Only two and a half percent of the water on Earth is fresh, with two thirds of that being frozen. Making its glaciers; which are receding at record rates all around the globe; our planet's leading source of fresh water, and many may completely disappear later this century. Well over a billion BBAs rely on Himalayan glaciers as their primary fresh water source. When they're melted, will those billions of Asians come knocking on our doors for a drink? At one hundred gallons per day, Americans use almost twenty times the amount of fresh water as the average 'third world' BBA. Yet unisex *waterless urinals* could save a few gallons of precious fresh water for every current wasted archaic toilet design urine flush reducing the severity of regional water shortages coming to a town near you later this century. Why haven't the BBAs in D.C. funded their development? Over sixty percent of our fresh water use goes to farming and ranching, half of that to irrigation. It takes twenty times more water to grow the same number of calories of beef as it does for rice. Already completely consumed, the Colorado River no longer reaches the Sea. When the global supply of un-synthesizable Phosphorus and the American grain belt's Ogallala Aquifer both run dry later this century, the 'world's breadbasket' could shift to the Eurasian Steppes. What impact may that have on the long term security of our sovereignty? Yet not one self-ignorant BBA in D.C. has the Geologic vision to start working on a NW Territories to Texas water pipeline; as all of those catastrophes looming later this century are as yet unseen by our myth infected BBA politicians who's myopic vision ends at their next election cycle; rather than at their inaction's impact on all our great grandchildren. Will they then explain to their own; *'but our economy needed oil!'*; when they have no water and can't farm the land? Geologic vision being the greatest strength of HS self-awareness, we must only elect Scientifically literate HSs who possess it in every election cycle forward; as The HS *perspective* now remains essential in preventing our species' extinction; and HSs trying to Reason with blind, myth infected BBAs equates to them trying to Reason with Chimps; no Geologic vision.

"Earth has its boundaries, but human stupidity is limitless." - Gustave Flaubert

Instead of funding those blind myth infected BBA's insane Simian aggressions for short term oil profits; perhaps we should be funding our HS Scientific Community with just a fraction of their *immense wasted military spending* toward mitigating all of those climate change catastrophes looming later this century? The fossil record shows our rapid increases in brain size coincided with; and so were likely our adaptive response to; periods of stressful climate instability. As during long periods of non-stressful stable climate; to which all our then small brained bipedal Hominin ancestors were already well adapted; we had no such increases in brain size. Perhaps we'll come out of this coming climate instability with even bigger brains? Will we finally use them? More recent tree ring histories show that extreme climate change also coincided with the demise of some of our species' most formidable archaic civilizations. Did we learn nothing?

'Red man build small fire, sit close; white man build big fire, sit far away.' Always stress conservation and recycling of every resource to all young children. Lest they be left wanting, let them waste not.

"Waste is worse than loss. The time is coming when every person who lays claim to ability will keep the question of waste before them constantly. The scope of thrift is limitless."
-Thomas A. Edison-

NOTES

SOLAR WISDOM

"It is not the most intellectual of the species that survives; it is not the strongest that survives; but the one that is best able to adapt and adjust to the changing environment…" - L.C. Megginson

Teach every BBA child this profound Truth; 'fossil fuel' is *fossil sunlight*. And that every week of current sunlight delivers more energy to the Earth than is stored in all the fossil sunlight deposits on this planet combined. And that Truth should always be their primary consideration in all of their life choices, from the site and orientation of their homes and businesses and even parked cars, to when and where they travel; to passively reduce their total energy needs by two thirds.

Our archaic ancestors understood this Truth and oriented their settlements to the Cardinal Points. Today's BBAs imagine some *'mystical'* significance; HSs know all our Solar Wise ancestors were simply utilizing all available Solar light and heat, allowing them to live well, even in the cooler temperate regions without burning fossil sunlight. Today's BBAs foolishly ignore this unlimited, free daily power source, which enslaves them into dependency of the ever more expensive fossil sunlight. Teach all young children they'll never be truly free, until their residence and vehicles are all energy self-sufficient; employing the latest Science in Solar electricity production and storage.

The 'internal combustion engine' was a 19th Century idea; has our species learned nothing since? Has Detroit been asleep for decades? Or receiving 'payola' from the Saudis? Earth's entire fossil sunlight supply is bought and sold several times over again every single day. We put BBAs on the Moon with our 1960s technology, yet Detroit can't build a 100+MPG terrestrial vehicle fifty years hence? Much of BBA produced hydrocarbon comes from autos, yet each BBA's total daily auto usage averages less than forty miles; well within the range of even 1990s electric car technology. Don't the Apes just love the fast shiny metal? They build their autos based on BBA ego and want; but they must now adapt to build their autos based on HS humility and actual need. Just because *'we can'*, doesn't mean that *'we should'*. There is never any reason for any BBA to ever drive faster than 70mph. If all autos were geared for a top speed of 70mph, we could double our domestic auto fleet's MPG overnight. Instead of the current chaos, HSs will standardize all speed limits at 70mph for freeways, 55mph for rural highways, 40mph for arterials and 25mph for all city surface streets. And standardize all autos sold in America to be geared to run at only 1200rpms at each of those 4 speeds. When any hybrid then exceeds 75mph, inertia shifts to its battery charging generator until it's back down to 70mph. And HSs will replace all asphalt with *'Solar Roadways'*, to power all our residential and retail energy needs, including our by then all-electric auto fleet. We'll then 'top off' our solely electric cars with our own Solar Home's batteries each night, finally freeing us of our soon to be exhausted, and ever more perilously destructive, fossil sunlight dependency forever.

As we'll obviously return to this unlimited power source once the fossil sunlight runs dry later this century, why not do it now, before any further alteration of our grandchildren's global atmosphere? Imagine what HSs could achieve if we spent just a fraction of the BBAs' *immense wasted military spending* on the research and development of improved Solar electricity production and storage technology, *'Smart Traffic Signals'*, and synchronizing auto gearing to standardized speed limits? Why not now? Who did you vote for? Chimps with iPads? Their time is over; this is HSs' time.

Our archaic Solar Wise ancestors would laugh if they could see BBAs' present predicament; *'Why pay for fossil sunlight, when the real thing comes up again every day?'* Until our species regains its lost Solar Wisdom, every BBA should go a full 24 hours once per week with no fossil sunlight.

" Almost every way we make electricity today, except for the emerging renewables and nuclear, puts out CO2. And so, what we're going to have to do at a global scale, is create a new system."
-Bill Gates-

NOTES

NUCLEAR WISDOM

"The greatest obstacle to discovery is not ignorance; it is the illusion of knowledge."- D. J. Boorstin

See the clever Big Brained Apes! See the clever Big Brained Apes get too big for their Ape britches! See the clever Big Brained Apes almost exterminate themselves and all the other species here too! See the clever Big Brained Apes 'split atoms' to release energy as intense as the surface of the Sun here in Earth's atmosphere before they fully understood that it would create a waste product that is deathly mutagenic for millennia to every species on this planet that comes anywhere near it! See the clever Big Brained Apes become drunk with excitement about their new found 'power', and lose their ability to make thoughtful long term judgments about its permanent lethality!

See the clever Big Brained Apes then imagine they could also 'contain' this surface of the Sun here on Earth, to boil water into steam, to turn turbines to generate electric 'power on demand'! See the clever Big Brained Apes' 'contained' surface of the Sun also generate that same waste product that is deathly mutagenic for millennia to any species on Earth that comes near it! See the first 25 year global death toll from Chernobyl's 'fallout' already exceed a million BBAs and continue on for millennia; while Fukushima's is now projected to be *'several magnitude'* worse!

"If we continue to develop our technology without Wisdom or prudence, our servant may prove to be our executioner."- Omar N. Bradley

See the clever Big Brained Apes then imagine they could just bury their permanently lethal waste product, so they just wouldn't have to look at it, and maybe it would all *'just go away'*! See it take more time to dig the repository for their lethal waste, than it does to fill it, but their Scientifically illiterate Rep. Chet Holifield didn't do the math! See the 'judgment' of *one single* unqualified Ape 'politician' endanger the welfare of *the billions* of us to come here ever since! See the clever Big Brained Apes all have great 'faith' in 'The State'; because they never lie, and they're always right!

"Honesty is the first chapter in the book of Wisdom."- Thomas Jefferson

See the honest Humane Sapients then realize that this permanently lethal waste product will never all *'just go away'*, and that *they should stop producing it in Earth's atmosphere as soon as possible* by switching their fission material to Liquid Fluoride Thorium; whose waste product is benign by comparison; used in 'rare earth' magnets and 'nuclear' medicine, but can't be used to create bombs! See that critical distinction, as in all the clever BBA 'faithful's entire history whenever *'us'* had any 'power' advantage over *'them'*, *'us'* always used it wherever possible to murder or enslave *'them'*!

"Knowledge shrinks as Wisdom grows."- Alfred North Whitehead

See the humbly Wise Humane Sapients realize we must adapt to that self-awareness derived HS Truth and is just one of the myriad of reasons why we must now 'remove' all Scientifically illiterate BBAs from all of our Legislatures and Judiciary, as their myth infections and lack of Geologic vision are destroying our Nation's Economy and Justice and Global Reputation! See them then replaced with Truth honoring HSs, whose only exports; instead of insane Simian aggressions; will be food, The HS Truth and MS Reactors; which can 'consume' the BBAs' abundant lethal waste while supplying all the world with power; even running Alvin Weinberg's visionary desalination plants to make all the Earth's coastal deserts bloom!

"We are drowning in information; while starving for Wisdom. The world henceforth will be run by synthesizers; people able to put together the right information at the right time, think critically about it, and make important choices Wisely."
-Edward O. Wilson-

NOTES

SCIENCE AND POLITICS

"Unthinking respect for authority is the greatest enemy of Truth." - Albert Einstein

Science and Politics are moral opposites; one seeks The Truth, the other advantage. Any Science corrupted by the BBAs' political process or profit motive cannot be trusted. BBAs are known to kill for money; do you think they're capable of lying for it? Teach your HS children; contrary to all HSs, many BBAs value profit above their own integrity. Many medical students state their motivation for becoming MDs as *'high income potential'*, not the desire to heal. The AMA and FDA are political, not Scientific organizations; they're not healers, they're profiteers, *and brilliant lobbyists;* as theirs is the only business in America enjoying Congressionally 'mandated', employer funded, guaranteed paying customers. *Pretty slick racket!* But you accept their recommendations on 'faith' at your own family's peril. As along with our deceived and abused Military, the cloistering of 'medical' students away from all rational outside peer review dissent may lead to dangerous isolation of whatever false conclusions their profit driven internal 'professional establishment' sometimes do draw from either illegitimate or misinterpreted research. And as their studies of new drugs and 'medical procedures', such as thalidomide and lobotomy, are all eventually tested on live subjects, that absence of outside peer review, combined with their huge guaranteed profit potential, and self-ignorance of *what they are,* and *where they came from,* and *how they got here,* often costs BBA lives. Which they claim is an *'acceptable risk of progress';* something you may disagree with if it happens to your loved ones. The AMA reports new multi-antibiotic resistant lethal 'bugs', yet their own studies show many more Americans are killed each year from *'medical misadventure'* at the hands of their own members; now the fifth leading cause of U.S. deaths; than by all known microbial pathogens combined.

Teach your HS children that their own highly evolved *homeostatic processes* are responsible for all *'miraculous healings',* despite the claims of the 'faithful'. If any MD asserts; *'It's a miracle!',* they're actually saying; *'I am ignorant of the causation and so I am ascribing it to a supernatural source.'* Teach your HS children to always ask their doctors; *'show me the Science';* and then do their own evaluation, as there's no substitute for independent research, and they only want to work with MDs who practice sound Science; not AMA witch doctors. Despite their educational 'credentials', MDs are just Big Brained Apes; who don't know anything they can't learn; and their MD training doesn't raise their IQ, as your child's own HS *perspective* does. So work only with self-aware HS MDs, by thoroughly researching each 'practitioner's background. Beyond Optimum Nutrition, their family's health challenges will best be solved by applying that MD training through the prism of only those self-aware HS MDs' humble reasoning, not the avarice of all the many profit driven, myth infected AMA 'practitioners'. How different would our Public Health Care 'product' be, if we nullified their exclusive guaranteed profit and directed research by openly public peer reviewed Science instead?

As the BBA genome is much shorter than many other mammals; if the technical obstacles to cloning have been overcome for one; they've been for all. It is only BBA hubris that imagines otherwise. As other Scientific advances have also outpaced BBAs' wisdom in their application, you must apply your own faithless, reasoned skepticism as to each advance's ethics and efficacy. Our species' irrational fear of death and love of profit has created a 'medical system' that spends more money on each patient in the last few months of their lives, than in all their previous years combined. Be constantly vigilant *against* any proposed 'medical procedure', simply because *'we can',* especially for the elderly; applying HS epistemological modesty every time. Caveat emptor! Every day, every BBA on this planet is endlessly bombarded with solicitations from those whose only goal is to separate them from their money, not to improve their family's health and longevity. Always do your own independent HS research. Trust no BBA with a political or profit motive.

"Than politics, the American citizen knows no higher profession- for it is the most lucrative."
-Alexis de Tocqueville-

NOTES

VACCINATION

"It is error alone that needs the support of government. Truth can stand by itself." - T. Jefferson

Vaccines are the only 'consumer product' in America with Congressional protections of anonymity and indemnity. So since 1986 they have quietly paid out millions in claims to families with vaccine injured children. While Big Pharma's annual global profits from State 'mandated' vaccines exceeds 30 billion; projected to triple within a decade; as pediatric offices' largest source of revenue is 'well baby' vaccinations. Because of government protections, our AMA vaccine profiteers don't have to disclose any data on harm to American children; but records made public in many other countries should make every prudent parent *question the safety and efficacy* of all those 'mandated' vaccines. As all those countries' documented disease 'outbreaks' show that the percentage of fully vaccinated children who contracted the disease often matched the percentage of unvaccinated children; proving that at least some of those vaccines don't provide the prophylactic protection their proponents claim, even with all 'booster shots.' And the claims those vaccines were responsible for eradicating diseases are specious, as their century long public health records show that many were only introduced at the very last stages of a pathogen's natural bloom and decline in their population; already diminished to just a few percent of their former prevalence; likely due to improved sanitation and other variables. And the first ever eleven cases of Autism described in the medical literature 70 years ago, have all now been linked to exposure to the newly invented potent toxin Ethylmercury, either in their home environments, or in their earliest vaccines in the form of their Thimerosal preservative; whose use in infants' vaccines has more than tripled since 1980; and is in Flu Shots too. The CDC 'mandated' only 10 vaccines before age 6 in 1983; today they 'mandate' 38. The incidence of Autism in 1983 was 1 in 6,000 children; today it is 1 in every 88. This *'unexplained'* increase has led many nations to reduce their 'mandated' vaccine protocols. Yet despite our American regulators admission of a causal link at their private 2000 Simpsonwood CDC conference we continue to expand ours, as in 2013 the CDC, the AAP, and the WHO still publicly assert that there is *'no link'* between Autism and Ethylmercury.

"It is difficult to get a man to understand something, when his salary depends on his not understanding it." - Upton Sinclair

Big Pharma claims double blind studies would be *'unethical'*, because they'd leave half our children *'unprotected'*; yet the AMA is already, in effect, conducting a 'vaccination experiment' on all of our children; while Congress quietly pays out millions to families with neurologically damaged kids. The new HPV vaccines have caused thousands of adverse reactions, some claim even deaths; and since 1974, every domestic case of Polio has been vaccine induced. If vaccines are based on sound Science; *let us review it.* Would Congress require this *secrecy* if vaccines didn't harm healthy kids? Read their product disclaimers; which of those myriad ingredients defeat the 'blood-brain barrier'? Every variable in Nature rides on a 'bell curve'. The safety and efficacy of each unique vaccine lies somewhere on it between the success of the original Smallpox vaccines on the one extreme, and the total failure of all HIV vaccines on the other. So evaluate every 'mandated' vaccine *individually,* and weigh the severity of that disease versus the known risks of vaccination. Prior BBAs endured many 'bugs' without any lasting damage, which now have 'mandated' vaccines. If you find yourself being bullied by anyone, find another 'practitioner'. Coercion and scare tactics aren't Science, and a parent has a Moral Duty to question the AMAs. So when your infant is just hours old, and they waltz into your birthing suite insisting that *'it's the law'* that they *'vaccinate your newborn'* against the sexually and I-V drug use transmitted Hepatitis; militantly show those AMA profiteers the door! Research all *vaccine adjuvants, SV40,* and the *Th17 Autoimmunity trigger;* mindful that Autism and Autoimmune disorders are unknown among the 300,000 Amish population; unless they were forcibly vaccinated.

"Autism rates are now increasing almost exponentially in proportion to vaccinations."
-Lloyd W. Phillips-

NOTES

CIRCUMCISION

"Orthodoxy means not thinking; not needing to think. Orthodoxy is unconsciousness."- G. Orwell

There is no more heinous form of child abuse than the non-consensual severing of a body part. Just as we will for all Catholic raped children; whose parents offered them up to their priests as lambs to slaughter; HS society will provide legal recourse for just such an abused child to receive redress from their criminally culpable parents upon reaching the age of adult consent. It is insane to 'believe' that 'God' ordered Bronze Age Sons of Abraham to mutilate their Sons' genitals eight days after birth, and it well exemplifies the danger and true depth of the barbaric depravity they're driven to by their delusively false Origin Story induced mental illness. And just how small must SOA's minds and genitals be to make them terrified of educated women? How small must they be to compel them to murder innocent girls simply for learning to read? Perhaps the SOA's homicidal misogynist cowardice results from overzealous circumcisions?

The intent of all patriarchal religions is the subjugation and abuse of woman and their children. Teach your HS children that they may judge the degeneracy and depravity of all BBA 'faiths' by the actions and practices of their male adherents; *'...but it's always been done this way...'*

"Mere precedent is a dangerous source of authority."- Andrew Jackson

Why are the SOAs so terrified of The Truth? If they courageously admitted they're evolved Apes, they couldn't still cowardly beat, rape, mutilate and murder their fallacious 'faith' subjugated and abused women and children could they? For 'faithful' fathers to prefer their children to be raped or murdered to admitting The Truth is mental illness manifested. Like 'The Method' and 'faith'; Dignity and depravity cannot coexist. *'Evil'* is the absence of a self-awareness derived empathetic Dignity; not the absence of an abusive 'faith'. While BBAs still practice depravities from our ancient past; HSs know for any of us to truly thrive, we must all thrive together in mutually honored Dignity. The SOA's child abuse and murder aside; have your children witness the greater SOA's culture, whose 'faith' constantly fosters raging crowds of angry males shaking their fists and chanting in hateful *mindless* unison. Why maintain a 'faith' that fosters their own endless anger and bigotry? What possible benefit does their own culture derive from their 'faith's hateful self-inflicted injury? Can your HS children see how their 'faith' mandating that half of their population be kept illiterate has retarded their Moral Evolution; while every other culture on Earth strives to eliminate illiteracy? Is it any wonder they're our global family's slowest 'special needs' siblings? Can your children even imagine *the relief* The Truth will finally bring to all those long suffering myth infected Apes? If not enlightened, they'll fall even further behind all the rest of us, once their oil's gone later this century.

Teach all children that any 'Science' not solely constructed upon The HS Truth of our Simian heritage must be rejected as spurious pseudoscience; as Science is a 'Method' of solely Rational thinking which rejects all lies and nonsense. Asked why he only owned several identical suits, Einstein replied he never wanted to *"waste any mental energy on trivial matters"*. Unlike the 'faithful' of Bronze Age Harry Potter novel 'Holy Books', and other past centuries' myths, HS Scientists know; *'the first is often the worst';* as 'The Method' continually improves each new version based on new evidence. It's simply unconscionable that the modern AMA has let ancient barbaric mythological rites to become the norm of their supposedly sound 'Science' based medical 'practices'. Seek Rational Truth honoring HS 'practitioners' for your Sons, not AMA witch doctors. A wise lifelong principle to teach all young children to adopt is that self-aware Humane Sapients never take advice from non-self-aware Big Brained Apes, no matter their 'credentials'.

"Still, instead of trusting what their own minds tell them, men have as a rule a weakness for trusting others who pretend to supernatural sources of knowledge."
-Arthur Schopenhauer-

NOTES

BREAST FEEDING

"Every time we liberate a woman, we liberate a man." - Margaret Mead

After subjugating and abusive patriarchal religions, and Ethylmercury preserved vaccines, the next great fraud and abuse perpetrated on confidence lacking, fearfully self-ignorant female BBAs is the 'commercialization' of the mother-child relationship by the male dominated for-profit corporations.

In addition to helping BBA mothers safely regain their prior weight by daily expending 500 calories; through many millions of years of Natural Selection, BBA breast milk has evolved to be the perfect exclusive food for all BBA babies for at least the first six months of their lives. Yet lying male BBA profiteers try to convince BBA mothers all over our planet that their substitute is a superior *'formula'* to Evolution's perfect food. This could only be possible within a universally self-ignorant species where the mother-child relationship is not held in the highest possible reverence, and the health and welfare of women and their children is valued secondary to corporate profits. Corporate profiteers from some of the industrialized west's most well respected companies even knowingly promote the consumption of their formula where their customers only source of mixing water is contaminated, giving their formula fed infants lethal diarrhea, which remains the leading cause of infant mortality globally. And as our species' fallopian tubes are not even fastened to our ovaries; a serious risk for 'third world' maternal mortality are ectopic pregnancies. Would a *'perfect'* God *'create'* us with such a fatal flaw? While Natural Selection can only 'work with' what has already evolved. Choose Truth.

What price will an innocent child pay for their non-self-aware mother's Evolutionary ignorance? A formula fed baby versus a baby that is breast fed for at least one full year, will have a weaker immune system throughout their entire life, resulting in much more illness and lost productivity. And even more importantly they will average five points lower IQ, which may equate to one full grade point level in school, and have a lifelong adverse impact on their true genetic potential. In addition to these avoidable limitations, studies also show that breastfed babies have much greater social cognition as well, including earlier language development and other interactive skills. Our own USDA has stated airborne Fluorides are; *"the most damaging pollutant to domestic animals worldwide"*. Now ranked only 174th, likely in part due to our Fluoridated tap water mixed formula, we now suffer much higher infant mortality rates than the exclusively breastfed areas of the third world who've successfully avoided all the west's depraved formula promoting BBAs. Breastfed babies have gotten along just fine for many millions of years without any lying male profiteers.

But ultimately the final responsibility lies with the mother as she must educate herself about her body's extraordinary Evolutionary adaptations; for she must use that knowledge to assertively resist all the fraudulent male profiteers' attempts to convince her of her need for their 'services'. The Evolutionary Truth empowers them with the high self-esteem and unshakable confidence they will need to embark on the greatest of all possible Humane Sapient journeys; Motherhood. And The HS Truth also eliminates their tolerance of any abuse from any 'partner' as well. Once self-aware they'll know that no HS male would ever abuse any female; so no HS Truth knowing female would ever tolerate the company of any BBA male who ever did. While currently a full third of all female BBA victims of homicide are murdered by their own BBA 'intimate partner'.

If a BBA can't commit to at least one full year of breast feeding, she should wait to have children until she can, as to not breastfeed may be as detrimental to her child's development as alcohol or drug abuse during pregnancy. Would BBA mothers so willingly surrender their infants to others' care while they chase other pursuits if they truly understood how crucial those initial years are?

"There are no illegitimate children, only illegitimate parents."
-Leon R. Yankwich-

NOTES

OWNER'S MANUAL

Would it ever be possible to arrive at a True conclusion when starting from a False premise? Is it possible to truly know 'who am I?' without first knowing The HS Truth about 'what am I?' Just as all ignorance based bigotry and conflict among BBAs is rooted in their self-ignorance; all of their other knowledge and wisdom is rooted in their True Enlightened Self-Knowledge. Their Ape body is the most extraordinary instrument our children will ever own, yet our public school system teaches them virtually nothing about what Science knows about it; leaving them to the mercy of all those who would profit from their ignorance, often with disastrous consequences. In addition to mandatory three year classes on growing and preparing meals of fresh whole foods; no child in America should graduate from Middle School without having studied a full three year A&P curriculum on their sublime Ape body; leaving the eighth grade with an up to date 'Pre-Med' A&P textbook; and a working understanding of their immune and endocrine systems and all their other brain and body chemical messengers which control their every predictable Ape behavior as to transcend the limitations of their Simian heritage BBA children must first understand them.

"To know thyself is the beginning of wisdom." - Socrates

If we teach all BBA adolescents what Science knows about the electrochemical anatomy and physiology of their thoughts and drives and emotions, they will become better informed, more well adjusted, and more compassionate HS adults. More able to cope with their own, and more understanding of others' predictable Ape behavior and actions; greatly reducing the epidemic of ignorance and fear based violence that teenagers currently inflict on themselves and each other.

Imagine a world where every child is taught The Evolutionary Truth about *what they are,* and *where they came from,* and *how they got here,* in a positive fearless environment. A world where adolescents understand their Ape hormonal drives are normal and healthy; and not *'evil';* but that with adult actions come adult responsibilities, and that adult consideration for the welfare of others before one's own is the foundation of The Humane Sapient Code of Ethics; as our Big Brain's self-awareness derived empathetic Dignity alone is what separates us from most other animals. A world where all children learn how to feed and care for their incredible Ape body to maintain its Optimum Health without myths or medications. And they learn that the average non-self-aware Roman barely lived to age forty. But by applying our species' new Empirical Knowledge of our Ape anatomy and physiology the average BBA's life span has recently increased by more than a decade per century; and if they all apply it their generation may well live longer than any other in our entire Evolution.

"One of the first duties of the physician is to educate the masses not to take medicine."
"The greater the ignorance the greater the dogmatism." - Sir William Osler

Our 'free market' BBA Media is saturated with solicitations from profiteers attempting to sell the self-ignorant worthless, or even dangerous products, designed to help them imagine they are not really just BBAs, but *'the unique eternal creation of the Divine Mind'*. But a self-aware HS Truth knowing adolescent cannot be victimized by their ignorance and will make judiciously healthier consumer choices. Including resisting the damaging effects of drugs and alcohol and unintended pregnancies. Early adolescent education of The HS Truth is *the key* to their informed resistance.

All BBA problems; from personal to international; result from *not knowing what they are.* Once they get that premise right, every solution becomes self-evident. How tragic that we fail to teach The HS Truth to all children, condemning them to live their entire singular life in fearful self-ignorance, only to die never once knowing the liberating wisdom derived from their own enlightened self-awareness.

"Without knowing what I am and why I am here, life is impossible."
-Leo Nikolaevich Toltoy-

NOTES

A BBA AND THEIR MONEY...

"To know how to grow old is the master-work of wisdom, and one of the most difficult chapters in the great art of living." - Henri Frederic Amiel

A dangerous new trend of the perfidious BBA Media are the mostly specious marketing claims of pharmaceutical and cosmetic product manufacturers direct to non-self-aware BBA consumers. The rapidly read voice overs, or tiny fine print disclosures of the alarming *'side effects'* of many medications and products are obviously casually dismissed by most non-self-aware BBAs to their great potential detriment, or they wouldn't continue to be marketed by that spurious method. In fact they couldn't sell them without the collaborative ignorance of the BBA public; they're counting on it. BBAs *who don't even know what they are;* and are more concerned with their outward appearance than with their inner health; should all be informed that they can live their entire lives with fungus yellowed toenails, and without a single eyelash; but they can't live a single day without their liver. And when warned; *'the side effects, though rare, may include blindness, or even death'*; hopefully only the kids who hadn't yet completed their compulsory 3 year A&P curriculum would be at risk. Botulism is one of the most lethal neurotoxins known, yet ignorant BBAs inject it into their faces? Remind your HS children that self-ignorant Romans once drank acidic wine from lead cups too.

"History does not repeat itself, but it does rhyme." - Mark Twain

BBAs now even market a hormone therapy designed to entirely shut down female menses; the imperative monthly cycle that has evolved over millions of years of 'selective' refinement to efficiently cleanse and restore the safe and healthy microbial balance of their reproductive system. No self-aware HS would ever even consider using a product of such extreme potential danger. Teach all your HS daughters well, because the negligently complicit BBA Media won't. The only part of BBAs' hair which is living is the root within the scalp. Can dead hair strands be made *'healthier'*, *'stronger'* or *'absorb'* anything without osmosis? One facial cream exclaims that their product contains; *'shitake mushroom complex'*; no wonder it's expensive! But the only way to improve one's complexion is through diet, not topical fungus. One toothpaste claims the citric acid in fruit *'weakens'* BBAs' tooth enamel; which has *co-evolved* to resists just such damage over many millions of years of Apes eating a primarily fruit based diet. Seems BBAs who can't explain why Cheetahs and Gazelles are the two fastest land mammals will 'buy into' almost any absurdity.

American BBAs spend billions annually on questionably effective, if not worthless, or even dangerous pharmaceuticals and cosmetics designed to perpetuate their delusion that they aren't really just BBAs; but rather something else entirely. It would be greatly amusing if it weren't so dangerous and wasteful. What if those resources were channeled into feeding starving children?

The majority of advertised personal product claims are provably fallacious, but as long as BBAs continue to value profit over Truth your children will remain at risk. Teach your children that all BBA Media is a form of hypnosis designed to manipulate them; always contrary to their best self-interest; and that avaricious BBAs will stop at nothing to separate them from their money. Teach them to examine every Media product claim through the prism of Evolution and 'The Method' to protect them from harm. Their HS *perspective* is their greatest strength and cannot be unlearned. Once self-aware, your empowered children will never view BBA Media the same way ever again; but until then, the BBA Media remains the greatest complicit obstacle to their HS Enlightenment.

"When you're young, you look at television and think there's a conspiracy. The networks have conspired to dumb us down. But when you get older, you realize that's not true. The networks are in business to give people exactly what they want."
-Steven Jobs-

NOTES

WILD APES

Wild Apes living in the natural habitat in which they evolved never suffer from any of the maladies currently plaguing and killing their city living close 'cousins' by the tens of millions. Obesity, CVD, HBP, cancer, diabetes, acid reflux, asthma and other autoimmune disorders, are all unknown in Wild Ape populations; despite them having essentially the same genetic makeup. How do BBAs respond to that Truth? They inject Wild Apes with cancer and 'study' how they die.

Two thirds of Apes' immune system is in the extraordinary mucosal lining of our gut, which other than our skin, has our Ape body's greatest interaction with the outside world passing through our hollow digestive system. Impervious to stomach acid strong enough to dissolve ingested meat, yet permeable enough to allow the absorption of nutrients and the elimination of wastes, rather than respect its extraordinarily delicate evolved chemistry, most BBAs assault it with atrocious diets. If BBAs want to live a long and healthy life they should 'study' how Wild Ape populations thrive, by eating what both our Ape bodies evolved eating, and exerting Wild Apes' same level of daily physical activity. And as pesticides are known endocrine disruptors and also damage Ape brains' Substantia Nigra, HSs grow as much of their own food as they possibly can, or know those who grow it for them at their local CSA. HSs also avoid eating anything that grows in BBAs' waste filled coastal waters, especially filter feeders, that are often served with our carcinogenic waste containing alimentary canals intact. Neither do HSs ever eat any mutagenic 'processed' meats.

"A person's time would be as well spent raising food as raising money to buy food." - F. H. Clark

To maintain Optimum Health; teach children to eat what all Apes evolved eating, and exert what Wild Apes still exert, as the greatest threat to their health and longevity is self-inflicted obesity. BBA obesity is low energy, low adventure, high maintenance, high cost, bad sex and early death. Teach them they are the product of millions of years of relentless environmental pressure creating their finely adapted body; the most extraordinary instrument they will ever own; and to never dishonor that heritage by scarring their sublime body with ink or piercings; *like bondo on a Ferrari.* Emulate the foresight of the 1960s NASA Moon Shot Scientists, not the 1560s Maori cannibals.

Teach good oral hygiene and to avoid Dentists, as excepting trauma care, as with the AMA, much of what the ADA asserts is profit driven fraud; as nothing they can put in their mouths is superior to their own evolved dentition. Fluoride; a toxic waste product of fertilizer production which the EPA has banned them from dumping into our air or surface water, yet claims is safe to 'dilute' into our drinking water; causes sarcomas in animal studies, and dental and skeletal fluorosis, as well as *"significantly lower"* IQ in children in multiple studies; with no studies proving cavity reduction.

Warn them they can't *purchase* good health. In the Natural World health isn't related to wealth or having health insurance, but only in the immoral BBA constructed legal fiction financial world. Some of the 'poorest' populations on the planet enjoy the most robustly healthy longevity, while some of the 'wealthiest' populations on Earth suffer the worst collective health and early death. All around our planet robustly healthy BBA longevity is solely correlated with how closely they still live to the soil and how fresh and raw are their primarily plant based diets. Ranked only 51st in longevity, a quarter of American BBAs now die of cancer, while a full third die of cardiovascular disease, despite all their many *'pharmaceuticals'* and *'medical procedures'* and *'health insurance'*. Poor health results from self-ignorance, not from poverty; as the fraudulent BBA profiteers claim; as there's only one Ape on this planet that has any control over what you put in your own mouth.

"Was the government to prescribe to us our medicine and diet,
our bodies would be in such keeping as our souls are now."
-Thomas Jefferson-

NOTES

SUGAR

Teach your Humane Sapient children that 'food' is the most powerful 'drug' they can consume. Every time they eat it triggers a blood infusion of the master hormone insulin, which creates a cascade effect of secretions from all their other endocrine glands, which control every function in their brain and body. Being their body's single 'hungriest' organ; their brain alone consumes more than a quarter of their total daily caloric intake. So no matter what they eat, their body must create some sugar (glucose) from it, as glucose is the only 'fuel' their Ape brain burns. Obviously their Ape body has evolved to be very efficient at this conversion process, so when they foolishly consume any refined sugar their blood glucose levels become dangerously high, and their pancreas gets exhausted secreting large amounts of insulin, which among many other things signals every cell to take in the blood glucose and convert it to stored intracellular fat. In time, with continued ingestion of refined sugar, their endocrine system, as well as all their other cells, become resistant to these abnormally high doses of insulin; resulting in all the devastating obesity related, life shortening diseases currently plaguing so many city Apes. Asked how it felt to be a billionaire, Bill Gates brilliantly replied; *"At some point, the food can't get any better."*

Refined sugar is a very recent addition to BBAs' Evolutionary diet. When it was first introduced to the royal families of Europe, it devastated many of their formally healthy members leading to their early deaths. While their poor peasants who continued to eat unrefined whole foods which take much longer to digest and therefore don't quickly flood their blood with glucose, for the first time ever began outliving their rulers, leaving far more offspring despite their poverty. 'Chips' and 'Soda' are now two of America's top five most consumed foods, yet neither are actually 'food', just sugar, salt and fat. Refined sugar calories now account for 10% of the American BBA diet. They're now literally eating themselves to death, all of their hospitals filled with sick and dying non-self-aware Apes. Most BBA death in this brief singular life prior to age 90 is now self-inflicted.

"I saw few die of hunger; of eating, a hundred thousand." - Benjamin Franklin

What is Sugar? Everything other than protein and fat. Carbohydrates and starches are simply long molecular 'chains' of sugar. When a plain baked potato hits your stomach acid, it's broken down into the equivalent of more than a cup of refined sugar. Think of refined sugar as *poison;* teach your HS children to never consume it. Teach them that fresh fruits and berries have co-evolved to taste just like candy to all seed dispersing mammals, if they never consume refined sugar.

The Inuit peoples of the Arctic Circle ate a diet consisting almost exclusively of meat, blubber and oil. Yet before European contact obesity and tooth decay were unknown among them, and robust longevity was their norm. But since Europeans introduced refined sugar and flours into their diet; tooth decay, obesity and diabetes became commonplace within just one generation.

The Pima Indians of the American Southwest eked out a living in another of the most barren environments on Earth. Earliest European accounts marveled at the incredible stamina of their lean muscular bodies, reportedly able to walk all day in the desert heat on just a few handfuls of local native forage. Tragically, since the European introduction of refined sugar and flours, these magnificent people now suffer prevalent middle aged diabetes and blindness and death. The Pima are now being studied to determine the genes that make their glucose conversion so efficient in the hope that they may help 'cure' diabetes; and their elders now teach their youth to return to their traditional diet and shun the European madness that now kills so many Americans.

"Man seeks to change the foods available in Nature to suit his tastes, thereby putting an end to the very essence of life contained in them."
-Sai Baba-

NOTES

OPTIMUM HEALTH

"A man is as old as his arteries." - Pierre J. G. Cabanis

Teach your children; obesity is dependency. It's about fitness, not 'image'. What fit HS would tolerate a dependent partner who can't even climb or scuba, and then leaves them widowed young? Another of the greatest frauds ever perpetrated against BBAs is the lie that we must pay a third party Ape to maintain our own Optimum Health. And what a testament to our species' credulous self-ignorance!

If your child needs to safely lose 1-3 pounds per week to reach their Optimum Health: have them eat like Wild Apes; as many richly colored plants, high in micronutrients as they want; hunger is a good sign; it means their body is burning up 'Fuel', stored and otherwise. Keep it burning - Eat! Their Ape body is a metabolic furnace; they must give it Oxygen and 'Fuel' to burn stored fat. And drink at least three quarts of nothing but pure water daily. Throughout the known Universe it is the sole elixir of life and every cell in their body has evolved to need it for proper function. Without adequate hydration their blood filter organs cannot function optimally, and they are the organs responsible for the metabolization, distribution and elimination of fats. Stay hydrated!

1. Consume no refined sugars, flours, or potatoes or any other starchy (sugar) tubers.
2. Consume no alcohol. (Ape livers metabolize alcohol by converting it into 'sugar'- see #1)
3. Consume no refined corn or other refined grain flours. Eat whole nuts and seeds instead.
4. Consume a daily minimum of three quarts of only pure Water. (Ounce for ounce sugared 'soda' is equivalent to beer without the 'buzz'. Think of it as 'liquid fat'; never consume it!)
5. Consume a daily minimum of 40 grams of plant fiber. (A single piece of fruit averages 3 grams; 1 cup of vegetables averages 4 grams; 1 cup of beans/legumes averages 14 grams)
6. And if they must; only consume small amounts of non-mammal meat as a 'condiment'. (Thin strips in a vegetable stir fry, or chopped up and sprinkled over a large salad with cashews)
7. Eat richly colored plants whenever hungry; never eat any 'food' that has its own commercial.

Have them get at least a full eight hours of sleep every single night, and walk as fast as they can without breaking into a jog on level ground for sixty minutes every single day.

Why should they walk instead of jog?

Their Ape body burns fat the fastest at seventy percent of their maximum heart rate, which is perfectly maintained by walking as fast as they can on level ground. They should be able to carry on a conversation at that heart rate, as this is aerobic exercise where they're feeding their metabolic furnace 'fuel' (stored fat) and oxygen. After losing much weight, they'll be tempted to start running, but don't, if weight loss is their goal rather than conditioning their heart and other muscles. They can always do that later after they have reached their Optimum Health, until then save their joints for all their many new lifelong physically active recreational travel adventures.

Why should they walk for at least one hour?

It takes their Ape body twenty minutes of walking to burn up the glucose and glycogen in their blood and liver, before it starts to convert and burn stored fat from their body's cells. If they walk for just twenty minutes after they start to convert and burn stored fat (forty minutes total), their meta-furnace continues to burn at higher than resting rate for an additional four hours after they stop walking. But if they walk for twice as long after they start to burn intracellular fat (sixty minutes total) their meta-furnace continues that higher burn rate for an additional twelve hours after they stop walking; three times as long. A geometric, rather than linear progression.

"We may find in the long run that tinned food is a deadlier weapon than the machine gun."
-George Orwell-

NOTES

BREAST CANCER

"The history of life on Earth has been a history of interaction between living things and their surroundings. To a large extent, the physical form and the habits of the Earth's vegetation and its animal life have been molded by the environment. Considering the whole span of Earthly time, the opposite effect, in which life actually modifies its surroundings, has been relatively slight."
"For the first time in the history of the world, every human being is now subjected to contact with dangerous chemicals, from the moment of conception until death." - Rachel Carson

Big Brained Ape males now suffer a 1 in 2 lifetime risk of cancer. Big Brained Ape females' risk is now 1 in 3, with Breast cancer being their leading cancer killer globally. And the incidence of breast cancer has been relentlessly increasing by almost two percent every year, with the very highest incidence increase taking place in the most rapidly industrializing parts of the world.

Other than the known city environment risk factors of increased water and airborne carcinogens, studies now strongly suggests that industrialized areas' *'light pollution'* may also substantially contribute to this dramatic breast cancer incidence increase as well. Because among all females studied globally, those that work shifts at night in Big Brained Apes' bright, artificially lit areas suffer *double* the breast cancer incidence rate compared to those that don't work shifts at night.

Artificial night lighting is a very recent addition to Big Brained Apes' Evolutionary environment. For almost four billion years all life on Earth has evolved with a reliable cycle of twelve hours of bright broad spectrum light, and twelve hours of total darkness in every twenty four hour period. It is only in this past century alone that BBAs have artificially altered that consistent daily cycle. That consistent daily light cycle had allowed our nervous system to evolve a means of sending signals about the time of day and the amount of light and darkness independent of our visual system. Light that comes into our eyes stimulates a non-visual pathway directly to the center of our brains. That information about light and darkness is there converted into a hormonal signal by the production of *melatonin,* which is then released to circulate into every cell in our body.

During the day when we are exposed to bright broad spectrum sunlight, BBA blood levels of melatonin are suppressed; but in the darkness of night we produce abundant amounts of it, and our blood levels of melatonin then rise dramatically. This evolved twenty four hour hormonal circadian rhythm is so sensitive, that it has been shown to be disrupted by BBAs' artificial night lighting, which can suppress our melatonin blood levels just as completely as bright sunlight.

Studies have shown that breast cancer tumor growth is dramatically increased in an artificial BBAs' night lighting suppressed melatonin blood level environment, while breast cancer tumor growth is actually retarded in a darkness induced high melatonin blood level environment. HSs anticipate that future research will confirm this definitive correlation between BBAs' artificial night lighting suppressed melatonin blood level and the expedited tumor growth in all cancers.

Teach your HS children to never work any shifts at night in BBAs' artificially lit environments. To maintain their Optimum Health they should respect their evolved light and darkness regulated hormonal circadian rhythm by always seeking the full healthy darkness of the country night, and avoiding the harmful artificially lit BBAs' city lighting after mid-evening. For as with all of their evolved Ape physiological traits, their very survival may depend on respecting it. If they cannot see the Milky Way each night, they should reassess where they're living this brief singular life.

"We may never understand illnesses such as cancer. In fact, we may never cure it. But an ounce of prevention is worth more than a million pounds of cure."
-David Agus-

NOTES

NICOTINE

All Big Brained Ape children learn by example; if you smoke, your children will smoke. Teach your children that nicotine is among the most addictive substances known to Science, despite all of the large tobacco company executives lining up with their right hands held high, one after another swearing before congress and the whole world that they didn't 'believe' it was. Lying BBA profiteers will do everything in their power to addict your children to their products. Contrary to all Humane Sapients, fascist Big Brained Apes value profit above their own integrity.

"... an American fascist is one, who in case of conflict, puts money and power ahead of human beings... there are undoubtedly several million fascists in the United States." - Henry A. Wallace

The tobacco alkaloid nicotine bonds to Ape brains' neurotransmitter receptors stimulating a very pleasant sense of mental clarity and well being, and to Ape body tissues' receptors stimulating a pleasant muscular tension, as well as stimulating or inhibiting many other systemic functions. A strong vasoconstrictor, it also restricts blood flow to all major organs, and nicotine addiction kills more BBAs every year than all other illicit and recreational drugs and alcohol combined. Each minute of exercise adds one minute to a BBA's lifespan, and each minute spent smoking subtracts one. Over a fifth of all BBAs still smoke, making it one of the leading contributors to premature BBA death. HS society will permanently ban the sale of all nicotine 'products'.

Clueless BBAs; *who don't even know what they are;* delude themselves that money thrown at research will 'cure' cancer, when The Truth of prevention is self-evident to HSs. With essentially the same DNA, Wild Apes never get cancer, unless we inject it into them to 'study' how they die. Maybe we should 'study' how they thrive cancer free in the wild instead? What do city Apes inhale and ingest and vaccinate for that Wild Apes don't? Research fundraisers *'walk sixty miles in three days',* when they should be walking sixty miles every month for the rest of their lives instead, and eating only what all Apes evolved eating and respecting their evolved hormonal circadian rhythm.

"Everyone should know that most cancer research is largely a fraud, and that the major cancer research organizations are derelict in their duties to the people who support them." - Linus Pauling

While nicotine itself is not carcinogenic, all of its natural tobacco delivery systems are. Until you quit, smoke the fewest number of high nicotine cigarettes you can, to feed your nicotine addiction with the least amount of lung irritating cancer causing smoke. Typical of lying BBA profiteers, they more heavily promote their far more profitable low nicotine brands, which use less of the expensive tobacco and so are consumed faster, making them much worse for nicotine addicts. Nicotine gums and patches don't help one quit because they don't eliminate the very drug you're addicted to, they only alter its delivery system, to delay the inevitable; and isn't that what you've been doing for a long time now already? You don't have a 'habit', you have one of the strongest Ape drug addictions known to Science. And you won't ever be free of it until you suffer through nicotine withdrawal for at least a full week, no exceptions. There is no easy way through it. You simply have to decide if you want to feel temporarily better for just the next few hours of your shortened sickly life, or feel permanently better for the rest of your longer robustly healthy life. Nicotine, not Marijuana, is BBA culture's 'gateway' drug. Teach children to avoid the complicit BBA Media glamorized, and always readily available nicotine. If they successfully avoid it early, they will be far less likely to ever smoke other illicit or recreational drugs throughout their lives. With nicotine, prevention is not only self-evident, it's indispensable. HSs' greatest obligation to their children's robust health and longevity is to convince them to avoid it when they're young.

"I never give them hell. I just tell The Truth, and they think it is hell."
-Harry S. Truman-

NOTES

THE BRAIN

Due to dramatic advances in medical imaging technology, HSs have learned more about the True electrochemical anatomy and physiology of how our brains actually function in just the past five years, than in all our prior years of collective research combined. One new insight is that our Ape brains are not yet fully developed until age 25. And the last area to develop fully is the pre-frontal cortex; which is the area responsible for our 'higher reasoning'; which 'thoughtfully restrains' our ignoble Simian impulses. This new insight will be the benchmark HS society will designate as the appropriate age of 'adult' consent and obligation; including any Military 'service'; as the primary Moral Duty of all parents is to protect their child until their own 'higher reasoning' has developed. What moral society would allow manipulation of its vulnerable youth by entrenched profiteers? The current average age of our culture's 'media consumers' is early teens. If we allow The State, Religions and Corporations to sell lies and nonsense to our teenage children in their BBA Media while they have *only half developed brains,* incapable of judicious rational thought, they'll never enjoy the Age of HS Enlightenment bring lasting Peace to our planet. Peace, Justice, Liberty and Self-Determination can only be attained by teen BBAs universal self-awareness of The HS Truth.

Another new insight is our brain's lifelong compensational 'plasticity'. The functions of damaged regions are now 'seen' to be taken over by neighboring regions. And we've learned that our brains even have the ability to grow new memory cells up until death; which can be stimulated by aerobic exercise and foods containing DHA. Researchers also now claim to have clear imaging 'proof' that both alcohol and drug addiction are indeed 'diseases', because they can now 'see' an actual physical change in the neural pathways of the addicts' brains. But with our new found understanding of our brain's 'plasticity', one questions whether a physical change still defines 'disease'? As they can also now 'see' that when those addicts do indeed, even with more difficulty, stop their substance abuse, their brains do, in short time, return to their original 'normal' function. Drug addicts will confirm; Oxycontin is Heroin; the most addictive opiate known to Science. Yet non-self-aware FDA Apes, approve non-self-aware AMA Apes, to prescribe it to non-self-aware BBAs, until they overdose.
Teach your self-aware HS children that all consciousness is now *observable* electrochemistry.
What their Ape Brain perceives as 'spiritual self' is actually an organic phenomenon.
The organic Ape Brain dreams the 'mind', the illusionary 'mind' dreams the 'soul'.

"Religion is all bunk." - Thomas A. Edison

As all experience is instructive; if an 'adult' contests this Truth, they may want to visit a nearby Wilderness Area with some trusted friends to 'monitor' them, and chew some peyote buttons, or ingest LSD, or consume psilocybin, as the archaic SOA rabbis were known to do to elicit their 'spiritual' visions; and convince themselves of this Truth through their own personal experience. But be forewarned, although their brain's functions will return to 'normal' as soon as those drugs leave their system, that single experience will likely permanently alter their 'worldview'; as the Fortune 100 companies, who pushed for the criminalization of LSD in the 1960s learned, when they discovered that when any of their young corporate recruits experimented with LSD, many would often soon thereafter abandoned their Ivy League corporate executive 'greed head' career track to pursue an entirely different path in life. Conclusive evidence of the true electrochemical source of all consciousness and those three similar compounds ability to help one understand that Truth, despite the BBAs' 'spirituality' obfuscation to the contrary. If every 'faithful' BBA on Earth understood that no Ape brain is capable of *'divination',* all our problems here would soon be over. For BBAs to transcend the limitations of our Simian heritage we must all first understand them.

"'You', your joys and your sorrows, your memories and your ambitions, your sense of personal identity and free will, are in fact no more than the behavior of a vast assembly of nerve cells..."
-Francis Crick-

NOTES

SEXUALITY

As more offspring is the simple unconscious and undirected singular result of Natural Selection, teach your teenagers that their own evolved procreation assuring 'drives' are normal and healthy. To teach teenagers; as the 'faithful' do; that sexual desire is *'evil'* is abuse, and may trigger crimes of unnatural repression, like their army of child raping priest deviants. Be vigilant; *conscienceless sociopaths* will often exhibit the following four traits in adolescence: lying; bed wetting; arson and cruelty to animals. Teach your teenagers; *'evil' is the absence of empathy;* not a healthy 'sex drive'. Teach them how to recognize their healthy hormonal Ape endocrine system 'drives', but that HSs take responsibility for their actions, and that BBA children raised without both parents are often at a significant economic disadvantage to those who are; so a lifelong commitment of support is required before any chance of pregnancy. Being unconstitutional; Humane Sapient legislators will nullify all State marriage 'laws', and employ civilly enforced 'parenting contracts' instead.

Lacking awareness of *what they are*, and *where they came from*, and *how they got here*, and therefore self-ignorant of the True organic electrochemical source of their consciousness; the 'faithful' all suffer from what Einstein referred to as *"a kind of optical delusion of our consciousness"*, sadly resulting in one of their more intense bigotries towards others whose sexual 'orientation' is contrary to their own. Their delusion of a 'spiritual' source of consciousness makes them imagine there is a 'moral choice' involved in the hormonally driven; and therefore pharmaceutically alterable; sexual orientation, when in fact, like melanin content, it's a genetic trait unrelated to either IQ or character, with no basis in 'morality' whatsoever. While self-aware HSs who understand The Evolutionary Truth are indifferent to others' sexual orientation, knowing it's irrelevant, and simply none of their business.

Admittedly, homosexuality is an intuitively contraindicated genetic expression, as Natural Selection drives only one objective; that being producing the highest number of offspring; and obviously that isn't achieved through 'same gender' pairing. So one would imagine it would be 'selected against' in every generation. Yet it persists in our gene pool in significant numbers. Homosexual relatives must then provide a 'selective advantage' in producing a higher number of successful *collective* offspring among BBAs' extended nuclear families; or it simply would not persist in our gene pool. Unless it results from a very common first generation mutation. The Bonobos, our closest genetic relatives, daily exercise opposite and same gender 'adult' pairing all troop wide. It's theorized that helps make their societies the very least aggressive, most 'intellectually cooperative', caring and compassionate of our Homininae relatives; resulting in more successful *collective* offspring. And don't the caring and compassionate deserve our understanding, rather than our fear and loathing? With almost half a million children in foster care; only half of whom ever even graduate from High School; America can ill afford to deny any caring and compassionate citizens of any 'orientation' their right to adopt.

"A society that condemns homosexuality harms itself." - Edward O. Wilson

Sodomy is unwise, as the BBA colon harbors one of the densest microbial communities on Earth; averaging 2 to 5 pounds of a BBA's total body weight. Not exactly a healthy genital 'playground'. Most BBAs would never consider eating raw sewage, but some of them witlessly choose to *ingest* that same 'microbial package' through an alternate transmission route. If BBAs insist on thrusting their genitals into a BBA sewer, they shouldn't be surprised when they get infected. With over four million American Hepatitis carriers, and an estimated 1 in 6 Americans carrying Herpes, and even higher numbers believed to have at least one STD at any given time, coitus is dangerous enough; without penetrating where Nature never intended. Leave that perilous activity to the self-ignorant.

"I have no objection to anyone's sex life, as long as they don't practice it in the street and frighten the horses."
-Oscar Wilde-

NOTES

ABORTION

Leviticus 17:11- *"For the soul of the flesh is in the blood..."*

To stop a beating heart is murder. Yet no female can ever be truly free, unless she, *and she alone,* controls her reproductive decisions. There's no Peace without Justice; no Justice without Liberty.

Teach all adolescent females that current medical imaging now proves that during the first three weeks after conception, the blastocyst's cellular division is extremely rapid, but it is still without a beating heart, and can therefore ethically be painlessly terminated before it reaches its potential future status. That window closes quickly however and cannot ethically be delayed, because the embryo's primitive two chamber heart begins to beat early in the fourth week after conception. After the potential child has a heartbeat, the only Scientifically ethical option for an advanced unintended pregnancy is adoption; as barely the size of a peanut shell, by just the ninth week after conception the fetus has a distinctly Ape face along with its beating heart. HS laws will reflect this Scientific Truth and stress the *moral urgency* of terminations; contrary to the current obscene BBA statutes that still allow for *dismembering tool* 'surgical' abortions well beyond the first trimester.

The Moral Evolution of any society can be judged by how they treat their most vulnerable members.

HS legislators will abolish all delayed dismembering tool 'surgical' abortions, and make RU486 or its equivalent 'morning after pill' universally available globally to any female of any age, without parental consent, who feels the need to terminate any pregnancy; for any reason; within that first three week window after conception, before the embryo's heart begins to beat; according to her own calculations alone, involving no other 'authority' than her own conscience alone; as to do otherwise would invite coercion, the antithesis of Liberty. As the Ethics are incumbent upon the timing; as with all great Liberty comes great responsibility. This Science based solution to one of BBAs' greatest 'ethical dilemmas' will require that every young adolescent female have completed an up to date 'reproduction' A&P curriculum by her age of menses; to empower her with the latest Scientific Truths available; so that she may make informed decisions in her own best self-interest, while also fully understanding and respecting the rapid status change of the cells growing within her driving the *urgency* of her decision. Every moral society owes this timely Scientific education to all of their young adolescent females, as they cannot possibly make ethical decisions without it. We must teach them The HS Truth and empower them with the means to act upon it responsibility.

"If problems are created through knowledge, then it is not through ignorance that we will solve them."- Isaac Asimov

Yet the BBAs who claim they oppose abortions are the very same anti-Truth zealots who oppose a timely adolescent female public Science education and public funding of free, universally available condoms and RU486; thereby causing millions more abortions themselves; and as such, instead of *'Pro-Life'*, we should henceforth refer to them as; *'the Anti-Truth Education Pro-Abortion Zealots'.*

"By denying Scientific principles, one may maintain any paradox."- Galileo Galilei

HSs won't try to Reason with BBAs who 'believe' in ghosts; we'd have better luck with Chimps. As with our deceived and abused Military, HSs won't allow myth infected BBAs to dictate their 'societal ethics'. The HS Truth can provide Rational solutions to all of BBAs' 'ethical dilemmas'. Teach your HS children that Humane Sapiens have the Moral Courage to apply them.

"Wisdom is found only in Truth."
-Johann Wolfgang von Goethe-

NOTES

GAY MARRIAGE

'Marriage' is an 'establishment' of deluded Catholic BBAs; nefarious Apes responsible for more repression, suffering and death of our species over their millennia long 'reign of terror' assault on The Truth than any other BBA institution. And whose last perfidious *'moral leader'* Pope Benedict even continued that insane assault into the 21st Century by forbidding his delusional 'faithful' from wearing condoms. Even in AIDS ravaged Africa, where infection rates can exceed twenty percent, and child starvation rates can be the highest on the planet. Thereby witlessly sowing the seeds of yet another Catholic genocide of the next generation of sub-Saharan Africans. He then refused to apologize for, or even acknowledge, his army of child raping priests; until he was shamed into it; and then that delusionally sick hypocrite Ape decreed *"same-sex marriage"* was; *"...today's most insidious and dangerous threat to the common good."* As opposed to his army of child rapists?

Yet LGBTs want to perform those mentally ill Catholic Apes' marriage 'establishment' because...? Have they lost their minds and self-respect? The only 'authority' that any BBA ever has over you is that which you grant to them. Why grant delusionally sick Apes *who don't even know what they are* the 'authority' to 'approve' of whom you love? Maybe our species is not yet worthy of Liberty and Self-Determination? Perhaps BBAs need yet another few centuries with a boot on their throats?

"Liberty will not descend to a people; a people must raise themselves to Liberty..." - Charles Colton

Those mentally ill Catholic Apes were the original fascists. From their torture loving Inquisition, to securing International Red Cross passports for dozens of Nazi war criminals to skulk away to South America along their 'ratline'; the Vatican's blood stained hands will never wash clean. Teach your HS children to run screaming from delusive Catholic dogma; for madness lies there. The American legal code holds that *'accessory after the fact'* bears equal moral guilt, and thus it bears equal punishment of the original crime. Harboring a known child rapist; as was the policy of every Catholic Dioceses; is the moral equivalent of raping that child yourself. Do the Catholic 'faithful' even know why Pope Benedict resigned? Don't they have a Moral Duty to find out? As their ongoing 'support' makes them *'accessories after the fact'* too. Occam's Razor asks; which is more likely; his army of child raping priests all suffered *'spiritual corruption'*? Or their delusive lack of self-awareness of The HS Truth of *what they are*, caused a lack of recognition of their own Ape endocrine system's evolved procreation assuring hormonal drives, leading to their twisting through years of unnatural repression into their BBA lust for raping children? That the Catholic 'Church' still even exists today is an alarming testament to the 'faithful's credulity and unwarranted trust in other BBAs abilities and intentions; as they remain the single greatest enemy of our species' free intellectual inquiry derived Enlightened Self-Awareness and resulting Moral Evolution.

Teach every young child that this brief singular life is short, much shorter even than they imagine, and to live it joyfully with the one they love by their side, for theirs is the only 'approval' they'll ever need. But seek it not from bigots, for they will always disappoint them. If there be a God, She surely must *delight in all love*, and despise the deludedly sick Catholic Apes' lies, odium, hypocrisy and murder. Secure legally binding mutual guardianship to ensure equal access of every kind, instead of wasting all your energies appealing to 'authorities' *that do not exist.* HSs will nullify all 1100 'State' marriage 'laws' as unconstitutional; as 'separating' Church and State our Constitution begins with: *"Congress* **shall make no law** *respecting an establishment of religion, or prohibiting the free exercise thereof;"* If only Jesus was returning; he'd throw out the Pope and priests and liquidate the Catholic Church's global assets, and use them to end child hunger and fund the free universal availability of condoms and RU486. Then Jesus would jubilantly perform every Gay Marriage ceremony requested of him.

"So long as the priest, that professional negator, slanderer and poisoner of life, is regarded as a superior type of human being, there cannot be any answer to the question: 'What is Truth?'"
-Friedrich Nietzsche-

NOTES

AIDS

The 'faithful' bigots claimed the global AIDS pandemic was *'God's punishment of homosexuals.'* Teach all young teenagers that Humane Sapients understand it's Natural Selection manifested.

"Nature never breaks its own laws."- Leonardo da Vinci

A great success of The Scientific Method in the 20th Century was the 1983 discovery of HIV as a blood borne infectious pathogen transmitted through exposure to infected blood or seminal fluid. By late 1984 that had been confirmed by peer review throughout the global Scientific Community; if one didn't want to catch HIV, one didn't have sex with, or share blood with, an infected person. With the exception of the Bayer Corp's mass murder by knowingly selling infected factor eight to hemophiliacs in Asia, why then have the vast majority of HIV contractions happened since 1984? Are we really such self-ignorant Apes, that we allow our hormones to rule our Reason? 60 million infected worldwide, with 7000 new contractions per day, a 1000 of those here in America per week, with an estimated 20% here remaining unaware that they're even infected. *'Adapt to environmental threats, or leave fewer offspring.'* Text book Natural Selection. Chock up another one for Darwin.

"Even if you are a minority of one, The Truth is The Truth."- Mohandas Gandhi

Teach all teenagers The Truth; there is no *'loving father'* in the sky who'll *'burn them for eternity'* later for not adapting now. Their 'punishment' will be metered out by Natural Selection in this life, by them contracting life altering Herpes, or Hepatitis, or HIV, or the yet coming even worse 'bugs'. Teach all teenagers The Truth; always use the organ between their ears, before they use the organ between their legs, or Natural Selection may cull them from the herd before they procreate too.

And they can't expect any help from the NIH; it's unlikely there will ever be an HIV vaccine, as it is among the fastest mutating viruses known to Science, with hundreds of distinct strains already identified. And the world's top immunologists collectively concluding by the mid 1990s that an effective vaccine would never be developed because this virus uniquely attacks the very pathway which they believe is the mechanism responsible for creating antigen immunity with other vaccines. And even when their experiments were designed to give their vaccines optimum advantage for success, no animal trials were ever successful. They no longer publicly state that fact; preferring now to say they *'remain hopeful'* that a vaccine will be developed someday so as not to lose their research funding; but tell your teenagers not to hold their breath while waiting.

Prevention remains the only viable solution, as Science already prescribed in 1984. Know their partner's entire sexual history, as they may still carry all of the pathogens that all of their partners collectively carried. They are quite literally having sex with everyone their partners had sex with. And teach your HS sons; female BBAs' leading complaint about their partners is that they come to bed unwashed, so to always honor their invitation into their bodies with healthful cleanliness. And to always use a condom; as they must live and die with the consequences of their choices.

The sixteen million orphaned children of the global AIDS pandemic are deserving of our greatest empathy and compassion. The HS world community will do everything in our power to minimize all their suffering. For contrary to the insane anti-condom dogma of the perfidious chief Catholic bigot, and the unethical mythology of the SOA, HSs do not punish innocent children for the 'sins' of their ignorant parents. Teaching The HS Truth is the most noble of all of our species' pursuits.

"It does tell us that given the currently available regimens that we have of anti-viral drugs, that we are not going to eliminate or eradicate this virus from individuals."
-Anthony Fauci-

NOTES

WWW

By taking advantage of their only half grown adolescent brains; to further their own rapacious interests; the 'broadcast' BBA Media have dishonorably manipulated your children's worldviews antagonistic to their own HS Enlightenment all their lives. Ask your children why, in our nation of over 300 million, do we only see the same 300 chattering Chimps on television? Is that fraction of BBAs capable of *'divining'* inerrant knowledge that all the rest of us can't? What qualifies them to even speak to your children? Kill their television; have them seek The HS Truth online. If you don't teach your children The HS Truth; who do you imagine will? Those that profit by deceiving them?

While adult BBAs remain clueless, their children are 'digital natives'. And HSs know the Internet is our field leveling game changer, and the primary tool in our endeavor to educate their kids out from under them toward finally achieving lasting global Peace in self-awareness derived Dignity and Reason in the coming new Age of HS Enlightenment. As all Scientific media ever produced in BBAs' entire history is already available online, if one simply knows where to look. No longer will the lying 'faithful' BBA parents be able to keep their children subjugated in abject ignorance.

Although our species' collective Empirical Knowledge, due in part to computerization, is currently estimated to double every decade, any HS Scientist will confirm, the Natural World, including our own bodies, is so incredibly complex that there are no true 'experts', only HSs more familiar with a study than the norm. As they could spend their entire lives studying just one discipline and only scratch its surface. *Perspective is everything.* Our body seems a Universe to microbes, our immune system a God. The more we learn, the more we learn there is yet to learn, making all the 'faithful's absolute assertions all the more laughably absurd with each passing decade. One must work very hard to remain ignorant of The HS Truth today with all the endless online educational resources available to them; *Good Effort!* The Jesus 'story' is a simple retelling of the Old Testament Isaac archetype, added to other earlier Bronze Age cultures' Astrological myths; Isis, Ra, El… IsRaEl.

"A Scientific Truth does not triumph by convincing its opponents and making them see the light; but rather because its opponents eventually die, and a new generation grows up that is familiar with it." - Maxwell Planck

Guide your children through the depth and breadth of the vast education potential of the Internet, but don't allow them to explore it ungoverned, or directed by their immature peers; for its proper utilization will deeply enrich their lives, but its naive misuse can be permanently destructive. So explain if any online 'service' claims to be 'free'; its *users* are 'the product' they then sell to others. To preserve their HS privacy teach them to always go online through a portal that doesn't capture their IP address, such as *'Startpage'*, as they must be constantly vigilant to protect their anonymity from online exploitation. Teach them the permanency of all digital media, which makes all their choices, good or bad, last forever. Despite their peers' foolishness, warn your children away from the *'mytwitface'* social networking sites. For unlike the 'broadcast' BBA Media, most kids haven't a clue of the capacities of just who, are tracking what permanent digital media at any time online.

Only after their completion of your guided web instruction, and after achieving an appropriate level of maturity, should you allow them to carry web enabled devices. Which will eventually become their constant educational resource. But hopefully then only used judiciously to access our huge, and ever growing organic body of evidence based Empirical Knowledge; for the first generation ever; now available online all around this planet to every Truth seeking BBA child.

"The sum of behavior is to retain one's own Dignity, without intruding upon the Liberties of others."
-Francis Bacon-

NOTES

TEENAGE DRIVERS
"Chance favors the prepared mind." - Louis Pasteur

Automobile accidents remain the leading cause of death for American teenagers. Before you allow any teenager to drive independently, have them ride along for a full day in the cab of a professional CDL trucker. So they can witness for themselves the utterly complete ignorance their average fellow BBA drivers have of the immutable Laws of Physics. Which if not for the constant vigilance of our nation's fleet of professional CDL drivers, would result in countless carnage on our roadways. That day will cement their understanding of the need for *'defensive situational awareness'*. And teach them that the most dangerous place for any car to 'linger' is beside an 18 wheeler, who may need that space to make a lifesaving evasion. So they should always pass any truck quickly, and then merge back in leaving plenty of safe space behind.

Applying the Laws of Physics, teach your children the usually fatal damage a speeding two ton car inflicts on impact with flesh and bone. Then ask them which is more deserving of their respect and consideration; the safety and welfare of a BBA pedestrian, or a painted stripe on asphalt? But teach them they should neither risk their own lives by swerving off the road to avoid hitting any other animal in their path, but should instead firmly grip the wheel and brake straight through it. Teach your children that power door locks are for security while parked and to never drive on the open road with them engaged, as emergency personnel can't unlock them following an accident where quick access may mean the difference between life and death. And have them always carry a seat belt cutting, car window breaking, emergency tool accessible from their driver's seat. And never place anything that obstructs their vision in front of their windshield or any side windows. Teach them front wheel drive is far safer than rear wheel drive. And the single most important rule of the road is to; *Always Keep Right* except to quickly pass. And whenever possible, never brake to merge, but rather always accelerate to speed and 'take' the lane. And if their car ever accelerates on its own, immediately shift it into neutral and turn the ignition off, then coast to a safe stop on the shoulder. And only the most ignorant of BBAs install ultra bright headlights, which will night blind their fellow Apes speeding two tons of steel and glass directly at them.

Today's autos are grossly over engineered for their intended purpose, as the weakest 'system' in any auto will always remain the Ape behind the wheel. So for BBAs to drive the equivalent complexity of spaceships terrestrially is absurdly inapt economically and environmentally.

"There is a spiritual hunger in the world today- and it cannot be satisfied by better cars on longer credit terms." - Adlai E. Stevenson

Five decades ago one could buy a brand new Detroit auto for just several months of the average American salary. Now BBAs will 'finance' their autos for several years, even though most will never use even a fraction of its capacity. HSs never 'finance' an auto, they buy them used at three years, then sell at six. Unlike BBAs, HSs don't drive their egos. They employ the quietest, most fuel efficient and most thoughtfully appropriate transportation for their specific needs each time. Capable of speeds far in excess of the highest posted limit, and weighing far in excess of what is necessary for the safety of its occupants, and embarrassingly energy inefficient, the Detroit fleet averaged better MPG after Carter's term than it did 25 years later. While clueless BBAs alter their grandchildren's atmosphere by squandering the finite carbon based fuels on needlessly faster cars, HSs' Geologic vision knows our 200 year *one time free pass* on cheap oil ends later this century, and that every gallon we now waste, is a gallon that won't be available for their real needs later.

"Infinite growth of material consumption in a finite world is an impossibility."
-E. F. Schumacher-

NOTES

FIREARMS

Teach all young children this universal historical Truth; free HSs own firearms, enslaved BBAs don't. Over 200 million disarmed civilian BBAs were murdered *by their own 'State'* in this past century alone. Teach them that only sociopathic despots and criminals fear an armed able populous. So the HS Code of Ethics dictates that every Truth and Liberty honoring able member of a free HS society has a Moral Duty to own, and be expertly skilled in, the use of firearms. As it is unethical to defer to others their Personal Moral Responsibility to defend themselves, their families, and our species' HS Liberties.

*"The Constitution of most of our states, and of The United States, assert that all power is **inherent** in The People; that they may exercise it by themselves; that it is their right and duty to be at all times armed..."* - T. Jefferson

*"The right of the people to keep and bear arms **shall not be infringed.** A well regulated militia composed of the body of the people, trained to arms, is the best and most natural defense of a free country."* - James Madison

"A system of licensing and registration is the perfect device to deny gun ownership to the bourgeoisie." - Lenin

"All we ask for is registration, just like we do for cars." - Charles Schumer

"Waiting periods and registration are only a step. The prohibition of private firearms is the goal." - Janet Reno

"If I could have gotten 51 votes in the Senate... for an outright ban... I would have done it." - Dianne Feinstein

"Guard with jealous attention the public Liberty. Suspect everyone who approaches that jewel. Unfortunately, nothing will preserve it but downright force. Whenever you give up that force, you are inevitably ruined. The great object is that every man be armed. Everyone who is able may have a gun." - Patrick Henry

"Among the many misdeeds of the British rule in India, history will look upon The Arms Act depriving a whole nation of arms as the blackest." "My nonviolence does not admit of running away from danger and leaving dear ones unprotected. Between violence and cowardly flight, I can only prefer violence to cowardice. I can no more preach nonviolence to a coward than I can tempt a blind man to enjoy healthy scenes." - Gandhi

"My religion is very simple. My religion is kindness." "But if someone has a gun and is trying to kill you, it would be reasonable to shoot back with your own gun." - Dalai Lama

"All political power comes from the barrel of a gun. The Communist party must command all guns, that way; no guns can ever be used to command the party." - Mao Tse Tung

"To disarm the people is the best and most effectual way to enslave them." - George Mason

*"When we got organized as a country, and we wrote a fairly radical Constitution, with a radical Bill of Rights, **giving (???)** a radical amount of individual freedom to Americans; it was assumed that Americans who had that freedom would use it responsibly... [now]... a lot of people say there is too much freedom. When personal freedom is being abused, you have to move to limit it."* - Bill Clinton

"Necessity is the plea for every infringement of human freedom. It is the argument of tyrants; it is the creed of slaves." - William Pitt

"To preserve Liberty, it is essential that the whole body of the people always possess arms, and be taught alike, especially when young, how to use them." - Richard Henry Lee

Teach them one may know a BBA enemy of HS Liberty by *'the company they keep'* and their cowardly lack of the Moral Courage to Trust their able fellow citizens with their **Inherent Right to self-defense.**

NOTES

CRIME AND PUNISHMENT

It is estimated that four percent; one in every twenty five; of all the BBAs that walk among us are conscienceless sociopaths, physiologically incapable of moral behavior. Conclusive medical imaging evidence of the unalterable neural pathways of their organic sociopathy indicate that those unevolved BBAs can never be rehabilitated, despite BBA society's long history of trying. Yet we allow the Scientifically illiterate 'State' BBAs in black robes; *who don't even know what they are;* and who can't even accurately define 'sociopathy', much less recognize its definitive imaging profile; to re-release those violent predators to repeatedly prey upon all the rest of us.

To protect our children, and all sociopaths' other assured future innocent victims, both inside and out of our *Ape Zoo* prisons, and to nullify the self-ignorance of our current sitting BBA judiciary, HS legislators will mandate that all *first offense* forensic convictions for Molestation or Murder or Rape results in permanent removal from our global BBA population. But HS society won't kill its own yet unevolved citizens; regardless of their actions; as The State 'collective' has no more moral right to murder than do its individual citizens. Especially not prejudicially, based on the convict's melanin content and 'income' level, as all *'us'* and *'them'* BBA societies have until now. One moral solution would be self-sufficient mid ocean *Banishment Islands,* lying between the tropics of Cancer and Capricorn, with permanent shelters and adequate fresh water, fishing nets, and hectares of stable well established fruit and nut tree orchards with corn and beans and rice. All violent 'medically proven' sociopaths convicted of those three crimes would be dropped onto the island of their gender, to freely live out their lives in a society of their own creation, with no outside world contact. Neither malice nor punishment; this *'catch and release'* violent sociopath program is preventative separation based on our current understanding of the immutable state of their organic sociopathy. BBA society's violent crime victim rates would plummet by banishing just those three worst types of unrehabilitatable sociopathic Ape predators to 'the islands' alone.

Our U.S. Ape Zoo population now numbers more than a quarter of all the BBAs imprisoned on Earth. Making the percentage of *'The Land of the Free'* now in jail exceed the percentage imprisoned in any other land, including the most *'repressive'* requiems; as one in every eighteen American males is now incarcerated; the very definition of a 'Police State'; having no Constitutional nor moral 'authority' to dictate what plants a citizen may choose to 'ingest'; so following Portugal's courageous 2001 example Liberty honoring HSs will decriminalize all unpatented botanicals, and free the full third of our Ape Zoo population convicted of such 'crimes'. We'll also deport all foreign national convicts, and banish all previously convicted *'third strike'* violent sociopaths. As until we HSs free every petty 'drug crime' convict, and banish all child raping BBA priests to 'the islands', we remain a profoundly sick society.

This decriminalization will do much to eliminate the huge, ever growing, and ever more violent sociopathic BBA international drug cartels; who are also often involved in trafficking guns and child sex slaves as well. And whom today at over four hundred billion in annual gross revenue currently control over eight percent of all international trade; now passing completely untaxed. DUI laws would then include all ingestable substances; yet many fewer Ape Zoos will be needed for the convictions of DUIs, lesser assaults and property crimes, as they'd all be greatly reduced due to The State controlled pricing, taxing and *local legal availability* of all of those substances. The unneeded Ape Zoos will be converted into homeless dorms with 'green' trade skill retraining and HS Truth education, including substance abuse prevention and addiction treatment programs. HS laws will then reflect the actual damage the criminal's actions do to the largest number of our society by jailing the sociopaths currently running our corrupt government and banking industry.

"Laws to suppress tend to strengthen what they would prohibit. This is the fine point on which all of the legal professions of history have based their job security."
-Frank Herbert-

NOTES

DUI

Every county Sheriff will confirm that only a tiny fraction of their county's population commits all of the violent crime; and they know who they are. They see the same sociopathic Apes cycle through 'the system' over and over again. And if they could just somehow 'remove' that tiny fraction from the population; they would have no violent crime. Are 'the islands' sounding more sensible to you now?

"Appeasers believe that if you keep on throwing steaks to a tiger, the tiger will become a vegetarian." - Heywood C. Broun

Due in part to the complicit BBA Media's glamorization; alcohol abuse is far more detrimental to our current BBA society, than the abuse of all the other 'controlled' substances combined; as the small fraction of anti-social BBAs among us continue to endanger all the rest of us with their self-ignorant choices, often leading to horrific carnage on our common roadways. Historically, the average DUI vehicular homicide victim has been killed by a *repeat* DUI offender on their *fifth* DUI offense. Is that really the best BBA society can do to protect our innocent children?

"One of the penalties for refusing to participate in politics is that you end up being governed by your inferiors." - Plato

HSs legislators will expand the 'definition' of 'Murder' to require banishment to 'the islands' for every DUI vehicular manslaughter, and repeal the current immoral laws absolving our non-self-aware BBA judiciary of legal accountability for their 'rulings'. HSs will hold all the current judiciary culpable by allowing the victims of all crimes committed by any 'released' convicts, to bring civil action against the judges who granted their release; requiring their 'removal' from the bench upon their adjudication. Replacing them all with Humane Sapient judges; as HSs have The Constitutional Right to be judged by our HS Truth honoring peers; imbued with the self-awareness derived HS *perspective;* rather than by the current Scientifically illiterate myth infected BBA judiciary *who don't even know what they are.*

"It isn't that they can't see the solution. It's that they can't see the problem." - G. K. Chesterton

HSs will then eliminate all future DUI fatalities by employing the proven Schick-Shadel anti-alcohol 'mammalian behavior modification'; because HSs fulfill all of their Moral Duties to their children.

First Offense: Sent directly to the Ape Zoo for a mandatory 10 day Schick-Shadel anti-alcohol shock therapy session; their license suspended for 60 days; and $1000 fine to release impounded vehicle, regardless of ownership; with vehicle sent to public auction within 10 days after offense if not paid. All proceeds used solely to fund further DUI 'mammal behavior modification' efforts.

Second Offense: Sent directly to the Ape Zoo for 3 mandatory 10 day Schick-Shadel anti-alcohol shock therapy sessions; their license suspended for 180 days; and $10,000 fine to release impounded vehicle, regardless of ownership; with vehicle sent to public auction within 10 days after offense if not paid. All proceeds used solely to fund further DUI 'mammal behavior modification' efforts.

Third Offense: Three strikes and they're out. Sent directly to the Ape Zoo to be detained until their trial. Upon their later forensic conviction, life in prison without the possibility of parole. Vehicle immediately sent to public auction upon third DUI arrest, regardless of ownership; before the insane BBA repeat DUI offender has any more chances to kill someone with it.

"Insanity is doing the same thing over and over again but expecting different results."
-Rita Mae Brown-

NOTES

POLICE

Humane Sapients *know what we are*, and *where we came from*, and *how we got here;* and that gives us a huge advantage over all the other Big Brained Apes *who don't*. We know what drives their thoughts and actions, and can usually confidently anticipate their behavior. While they are often *dependently confused and fearful,* much of this life remaining a complete mystery to them.

"They are not in control; we are." - Mohandas Gandhi

Teach your HS children; never ever speak to the BBA Police without competent Legal Counsel present. Never say anything to the Big Brained Ape Police except; *'I invoke my right to remain silent, and I want an attorney.'* Have all your Humane Sapient children memorize that statement. Any competent attorney will confirm this Truth. There is no possible benefit to talking to Police, no upside potential, none. While there is currently endless downside risk, potentially even their execution. Teach all your HS children that no matter what mistake they may have made, or what misunderstanding they may be involved in, it can be sorted out and corrected in time by always telling The Truth to you, their parent, and to their competent Legal Counsel. But until that time, they should always remain completely silent if they ever find themselves in Police custody.

"The greatest friend of Truth is time, her greatest enemy is prejudice, and her constant companion is humility." - Charles Colton

Our courageous men and women of *'the thin blue line'* perform the most dangerous task on Earth. A third of all Police killed on duty are murdered by sociopathic gang members. So unavoidably they develop an *'us'* and *'them'* adversarial perception of all the rest of BBA society along with all the sociopaths and other least evolved BBAs among us that they must 'deal with' every day. So out of necessity they have developed techniques of verbal coercion so sophisticated that the average self-ignorant BBA in custody doesn't even recognize when they're being manipulated.

"If you don't know who 'the sucker' at the card table is, it's you." - Old Gambler's Saying

Always remember, it's legal for them to lie to you, but it's a crime for you to lie to them. As with Martha Stewart and Amanda Knox, unrepresented BBAs often talk themselves into convictions. They aren't kidding when they say; *"Anything you say can and will be used against you..."* Every honest citizen breaks laws on the books every single day without even knowing it. The State designs their Enforcement Codes as such so that any suspecting Police Officer can find 'probable cause' to 'detain'; *which is an 'arrest';* almost anyone for 'questioning'.

"I am not a friend to a very energetic government. It is always oppressive." - Thomas Jefferson

Our 'government' was designed to be an unobtrusive servant in our lives; fix the roads, take out the trash; otherwise seen but not heard. But the Police exist to protect The State's control; not to defend your Civil Liberties. They have no moral authority to restrain your free intellectual inquiry; but only how your actions impact others. Anytime you invite the Police into your life, you surrender Liberties and control to The State. Contrary to the often dependently confused and fearful non-self-aware Big Brained Apes; independently self-sufficient, self-aware Humane Sapients never need dial 911. All the 'government' any Humane Sapient will ever need is between their own ears. The BBA State has proven they can't even 'govern' themselves; do they really imagine we'd allow them to 'govern' us?

"I would rather be exposed to the inconveniences attending too much Liberty, than to those attending too small a degree of it."
-Thomas Jefferson-

NOTES

PRISON

"Educate your children to self-control, to the habit of holding passion and prejudice and evil tendencies subject to an upright and reasoning will, and you will have done much to abolish misery from their future and crimes from society." - Benjamin Franklin

Teach all young children, if Evolution is True; which all Empirical Evidence from every Scientific discipline without any contradictions certainly suggests; than this singular life is the only time in all the Eons of time and space, that the consciousness each of us knows as 'self' will actually exist. Most self-ignorant BBAs sleepwalk through this brief, extraordinary singular life experience; only a tiny fraction are fully awake self-aware HSs who exist in a perpetual state of *awed astonishment.* Ask your HS children; just how profoundly self-ignorant does a BBA have to be to choose to spend this preciously limited, extraordinary singular life experience in an 8 by 12 foot Ape Zoo cage?

"What is necessary to change a person is to change his awareness of himself." - Abraham Maslow

BBAs' socializing conscience is 'selected for' by more offspring, but some aren't there yet, making much of our Ape Zoo population unevolved sociopaths, organically incapable of moral behavior. But many there are not sociopaths, and are imprisoned just because of their poor choices. Some may be excused due to mental illness, but many of the others made those poor choices simply because *they don't know what they are,* or *where they came from,* or *how they got here;* and so they suffer from the 'stupidity of ignorance'. Both etiologies will often result in multiple criminal incidents, so HS Legislators will mandate that every BBA arrested for committing a violent crime must submit their DNA, and must also submit to a sociopathy determining imaging brain scan.

Teach your judiciously self-aware HS children to never ever resist the Police. No matter what mistake they may have made, or what misunderstanding they may be involved in, resisting the BBA Police will only compound the situation to their severe detriment. Teach them to always immediately comply with all BBA Police commands. Teach them that instead of resisting the Police, they must resist *their own* Ape endocrine system's hard wired 'fight or flight' adrenal response to perceived threat, evolved over millions of years of environmental threat survival.

"I will face my fear. I will permit it to pass over me and through me. And when it has gone past I will turn the inner eye to see its path. Where the fear has gone there will be nothing. Only I will remain." - Frank Herbert

Teach them to always remember that the Police are just Apes too, with that exact same evolved endocrine system's hard wired adrenal response to perceived threat. So a calm HS never excites a BBA with gun who has trained to make *instant decisions* in response to that perceived threat. For their own safety, teach your children to never incite an adrenal response in the BBA Police. Teach them they cannot outrun a radio, so they must defeat the sudden adrenalin burst that tells them to try; because no 'flight' ever ends any other way than an arrest with additional charges; so any 'attempted flight' only adds more time confined in that 8 by 12 foot Ape Zoo cage later. Teach your HS children that in any interaction with the excitable BBA Police, calm HS humility will always serve them better than righteous indignation. Their very life may even depend on it.

Would non-self-aware Big Brained Ape inner City Kids still murder each other in such great numbers, if we taught them all the Empirical Evidence derived HS Truth *perspective* instead?

"More police, prosecutors, three strikes and mandatory sentencing laws, the death penalty, and putting nearly a million blacks behind bars have done little to curb the black-on-black carnage."
-Earl Ofari Hutchinson-

NOTES

CITY KIDS

"It is no measure of health to be well adjusted to a profoundly sick society." - Jiddu Kirshnamurti

Young Big Brained Apes that are born and raised in our larger *Ape Hives* are dying of profound self-ignorance. Entire neighborhoods of inner Ape Hive children graduate from Middle School directly into our Ape Zoo population because they literally can't conceive of any different life. When children are raised in a perversely self-ignorant hell, odds are they'll grow into demons.

"A person hears only what they understand." - Johann Wolfgang von Goethe

HSs give them the same advice we give to those living near violent 'faithful' sociopaths; *'Move!'* But it falls on deaf ears because they are wholly unaware of any other possible life, and America cannot afford to continue to lose their potential, cut down in adolescence in such great numbers. It is immoral for America to continue to allow these vulnerable young BBAs' testosterone and adrenalin and self-ignorance to control their destinies; without giving them a fighting chance. Poverty does not create social problems; self-ignorance based social problems create poverty.

If the violent are proven sociopathic, send them to 'the islands'; if they're mentally ill, treat them; if they're simply self-ignorant, educate them; by teaching all of them The Humane Sapient Truth. It won't be easy, because like the now record numbers of BBA soldiers subjected to unrelenting violence and terror and lies, their Ape psyches have been shattered, and many suffer from PTSD. So in their extreme case, instead of taking The Truth to them, we must take them to The Truth, through an intense total immersion into the Natural World where The Truth abounds undeniably. Change their environment, to change their minds, to change their chances, to change their lives.

"Come forth into the light of things, let Nature be your teacher." - William Wordsworth

All HS Public School systems will send every inner Ape Hive child to a National Monument or Park in the summer after the fifth grade for two months of intensive Evolution Science and trade skill education applied in maintenance and repair work on the Park's facilities, roads and trails. The valuable service work they'd provide would more than offset the cost of this HS program, although its true focus would be on instilling a sense of possibility and choice in the hearts and minds of these vulnerable young BBAs who have grown up their entire lives in perversely self-ignorant neighborhoods without either, where the sense of hopelessness and despair is palpable.

"Until we're educating every kid in a fantastic way, until every inner city is cleaned up, there is no shortage of things to do." - Bill Gates

And when they return home with hearts and minds now fully open to this life's true promise and previously unseen opportunities, they'd all say; *'Mom, Dad, we've got to get out of here!'* Then together begin a course of action resulting in their timely relocation to 'greener pastures'. Once they know the self-aware HS Truth, the solutions to all their problems become self-evident. If you live in hell, best advice is to get yourself and your parents out of there as soon as possible. How can we expect City Kids with only half grown brains to make decisions in their own best self-interest when The State, Religions and Corporations all lie to them every day in their BBA Media? Let's fill all their young Hiver heads with the enlightened self-aware HS Truth *perspective* instead. Give Humane Sapient Teachers just sixty days out in the Natural World classroom with every non-self-aware inner Ape Hive Big Brained Ape fifth grader, and we will change this world!

"A man's mind stretched by a new idea can never go back to its original dimensions."
-Oliver Wendell Holmes-

NOTES

APE HIVES

"Why should we tolerate a diet of weak poisons, a home in insipid surroundings, a circle of acquaintances who are not quite our enemies, the noise of motors with just enough relief to prevent insanity? Who would want to live in a world which is just not quite fatal?" - R. Carson

The Industrial Revolution required clustering large groups of factory laborers closely together. Global history has seen the colonization and then later abandonment of new territories after they became despoiled. Yet after the Industrial Revolution's centralized power and plumbing, BBAs remained clustered in those poisonously despoiled areas long after necessity. There's no current or future need for Ape Hives today. The Ape Hives have simply become *clusters of dependency*. Don't Hivers yearn to breathe free in the wide open spaces of this beautiful planet, where the air and water and views are so clean and pure and vast; rather than live, and breathe in, other BBAs poisonous filth? Beyond the stench, Hive living so dulls Apes' senses, they become lulled into a potentially lethal false sense of security from ignorance of their dependency based vulnerability. How many have ever considered what they'd do if food stopped arriving at their Hive? How many have ever once made a plan? One suffers a vain credulity to buy into Joseph Smith's absurd tales, but one must admire Mormons' *'disaster preparedness'*; as each and every family is charged with stockpiling a full two year supply of food and water; and the means to defend it; come what may.

"I view great cities as pestilential to the morals, the health and the Liberties of man. True, they may nourish some of the elegant arts; but the useful ones can thrive elsewhere; and the less perfection in the others, with more health, virtue and freedom, would be my choice." - Thomas Jefferson

With business meetings now online, there will be little need for business travel, and no need for Hives, as factories will be uniformly spread across the land in smaller new self-sufficient towns. Detroit foretells the future, the useless Ape Hives will continue to decay and then be abandoned. While current Hivers continue to spend 8 hours of this singular life experience in an 8 by 12 foot cubicle watching a small flat screen, to afford them the privilege of spending another 8 hours in a larger cubicle watching a bigger flat screen, then sleep for 8 hours, only to wake and do it all over again. How is that any different than spending all of their lives in an 8 by 12 foot Ape Zoo cage? As the ground beneath our feet is always moving, no HS would consider sleeping or working in a higher than three story structure. Yet non-self-aware Hivers daily pack into towers oblivious to the potential danger. One WTC tower 'survivor' tells of everyone on his floor 'packing' the stairs trying to exit after the second plane impacted, when a 'security guard' announced that; *'the safest place for you to be is back at your work stations'*, whereby every BBA but the survivor, dutifully returned like lemmings to the cubicles of their death. He alone ignored that advice and exited his tower before they were *'pulled'*, leaving him the sole survivor of his floor. Just how dulled does an Ape's senses have to be to sheepishly accept life or death advice from a 'security guard'? They'd all been 'dead' in their cubicles for years already. Run screaming from your Ape Hives to explore this extraordinary planet before you wake up one day to find your life over before you've even 'lived'.

"Twenty years from now you will be more disappointed by the things that you didn't do than by the ones you did do... Sail away from the safe harbor... Explore. Dream. Discover." - Mark Twain

HSs know all experience is instructive, even if it kills you; being then still instructive to others. Ask your HS children; what lessons should the BBA Hivers have learned from that experience?

"I think our governments will remain virtuous for many centuries; as long as they are chiefly agricultural; and this will be as long as there shall be vacant lands in any part of America. When they get piled upon one another in large cities, as in Europe, they will become corrupt as in Europe."
-Thomas Jefferson-

NOTES

WEALTH

"The more liberated one feels the less one needs." - Henry Miller

Teach your Humane Sapient children that the only true value of 'wealth' is unrestricted mobility. HSs who understand the Truth of Evolution and the brevity of this singular life, rather than the BBA fantasy of *'eternal awareness'*, realize that Natural World exploration and adventure travel are this life's most enlightening pursuits, and sharing both with loved ones are its greatest joy. The sole reason to achieve 'financial independence' is to enable one to travel freely at will; and given this life's brevity; the younger and healthier any BBA achieves that status the better. Never *'rich and famous'*, but rather *'rich and anonymous'*, as *'fame'* is its own self-inflicted bondage. The FBI attributes much success in their criminal capture to the fact that the average BBA lives their entire life within a fifty mile radius; reflective of the pathetic limit of the typical BBA's curiosity. BBAs are just livestock farmed for their work product by The State, Religions and Corporations.

"Wake up! To see the farm is to leave it." - Stefan Molyneux

Like Native Americans, HSs know that material 'possessions' are an illusion, a BBA legal fiction. Read your 'Deed of Trust', you don't 'own' anything. 'The State' only allows you to maintain that illusion as long as it's profitable for them to do so. You cannot own what can be taken from you. *Never 'hold' possessions in your own name, but rather in trusts.* You don't own your possessions, they own you; as any possession that ties any BBA down to one place in time and space is a self-imposed imprisonment. The only thing you can truly own in this brief life is your Intellectual and Moral Integrity. The Europeans were compelled to exterminate all the Native Americans; not just for their lands; but because they knew that they were all truly free men, and that they themselves were not, and that as long as they lived as free men, they would remind them of that; and that was intolerable to them. Never underestimate the 'power' of simple petty Ape envy and jealousy. Until the U.S. honors all our Native American treaties, we remain an immoral and self-debased people.

One must travel to understand the value of traveling. Unrestricted mobility is the sole measure of True freedom. Truly unrestricted mobility freedom can only be achieved in complete privacy. Teach your HS children to protect their privacy as if their very life depends on it; because it does. The cowardly BBA State is the antithesis of True Liberty; always disobey their 'laws' that invade and restrict your HS privacy and mobility at all times. Privacy is an Inherent Right; *not a crime.*

"The people never give up their Liberties but under some delusion." - Edmund Burke

Teach your children to *'pack lightly'* on their journey through this brief singular life, as any well traveled HS will affirm, if you can't carry it with you, you don't really need it. HSs live in the here and now; only sleepwalking BBAs dreaming of the *'here after'* imagine the need for 'storage units'. If your home burned to the ground today, what possessions would you truly need to reacquire? A great annual HS exercise in our world of shrinking space and exploding technological innovation involves reviewing every personal 'possession' with just two questions; *'Have I used it in the past year? Am I likely to use it in the coming year?'* If the answer to both questions is *'no'*, then recycle it to another HS that will make use of it. And HS couples know 'shiny rocks' are the default gift of duped livestock BBAs; devoid of both curiosity and imagination; and instead use their *'two months' salary'* for an extended travel adventure every year of their shared life journey. Teach your children to 'spend' their finite life resources wisely, and to scrutinize the true 'cost' of every 'possession'.

"Most of the luxuries and many of the so-called comforts of life are not only not indispensable, but positive hindrances to the elevation of mankind."
-Henry David Thoreau-

NOTES

THE STATE

America was an idea; not a place. The foundation of our original Constitutional Republic was the Courage to Trust one another with Liberty and Self-Determination. Sadly our current D.C. 'State' is a collection of criminally avaricious, non-self-aware, cowardly BBAs, who don't exhibit our Founders' Moral Courage to Trust 'The People'. If they don't trust all of us with Liberty; then they can't be trusted with 'authority'. Liberty honoring HSs have then a Moral Duty to resist; by the courageous examples of Gandhi and King; any intrusion that treasonously despotic collective attempts to insinuate into our lives. More and more our current avaricious D.C. State does *to* The People, rather than *for* The People. While lacking even the simple rational judgment to determine The Truth of *what they are,* and *where they came from,* and *how they got here,* those myth infected D.C. Apes yet scoff at The Constitution by presuming the 'authority' to dictate our own lives to all the rest of us? Just profoundly self-ignorant? Or sociopathic?

"A good person will resist an evil system with his whole soul. Disobedience of the laws of an evil State is therefore a duty." - Mohandas Gandhi

A juvenile 'faith' in The State is an extension of the same in religion. If you teach BBA children of a benevolent 'parent' in the sky, they'll imagine one in government too. An HS never cedes any 'authority' to the BBA State collective's chattering Chimps in suits and black robes. Nor ever asks for, nor accepts any assistance of any kind from same, as HSs are self-sufficiently independent of the cowardly BBA State in every way. The Natural World is the only Truth in this life, as the BBAs' created legal world is a fictitious illusion, designed to enslave the dependently confused and fearful. Teach your HS children that as long as our myth infected State collective *don't know what they are,* they will fearfully presume the 'authority' to control them; so lest they be complicit in their criminal foreign wars for profit; they must refuse all commerce with the BBA State, while freely ignoring all unconstitutional laws restricting their Civil Liberties; by instead always acting in compliance with the HS Code of Ethics; for The Objective Truth is the only genuine Moral Authority in this life.

"Never do anything against conscience even if The State demands it." - Albert Einstein

The Scientific Method proves that the worst possible justification for continuing any status quo is; like religious adherence; the perniciously ignorant claim; *'...but it's always been done this way...'* Only if the BBA State collective grows the Intellectual Integrity and Moral Courage to admit The HS Truth; that all of their constructs are merely legal fiction creations of BBAs, bearing no *'Divine Providence';* and as such are always subject to HS peer review as to their benefit or detriment to the advancement of our species' Moral Evolution, with all of their construct's nullification on the table if determined appropriate; shall any HS ever be morally obligated to honor any one of them. Their admission will begin with those treasonously despotic D.C. Apes' nullification of every oligarchic law and 'pension' unique to themselves; the cessation of all their criminal foreign wars for profit and illegal domestic oppression; and the full restoration of our original Constitutional Republic.

"How does it become a man to behave towards the American government today? I answer, that he cannot without disgrace be associated with it." - Henry David Thoreau

Whenever you hear the words *'the government',* replace them with the words *'my neighbors';* only then may you begin to understand True Liberty. While the current cowardly BBA State collective deludedly debate *what is,* teach all Truth honoring HS children to courageously ponder *what could be* in the restored Moral America; as they obviously can't leave that to the sociopathic Apes in D.C.

"It is not the function of our government to keep the citizen from falling into error; it is the function of the citizen to keep the government from falling into error."
-Robert H. Jackson-

NOTES

DECEPTION

*"The right of The People to be secure in their persons, houses, papers, and effects, against unreasonable searches and seizures, shall not be violated, and no warrants shall issue, **but upon probable cause,** supported by oath or affirmation, and particularly describing the place to be searched, and the persons or things to be seized."*- Amendment IV; American Bill of Rights.

What American HS would have ever imagined as they boldly set out in a horse drawn wagon to cross an entire continent of unknown threat filled wilderness with a long experiential distrust of 'government', that just two hundred years later their own absolutely innocent descendents would cower in such terror of Apes living in caves on the complete opposite side of the planet, that they would line up like cattle to allow agents of The State to radiate, fondle and strip search them out of an *unconstitutional presumption* of their guilty intentions to do harm to their fellow citizens? What American HS would have ever imagined as they jumped out of a landing craft directly into a hailstorm of bullets on the French coast while saving the world from tyranny, that within their lifetime their progeny would have to *endure* America's national news anchors quivering in terror for all our delightedly amused enemies to see, while asking agents of The State in their shaking voices; '...Are we safe?...Can you guarantee our safety?...'; because *one* third world BBA cave dweller out of *ten million* annual U.S. flights failed to ignite his 'shoe bomb'? *Perspective?*

What American HS would have ever imagined that they'd be asked to 'believe' what they are told by the BBA Media; instead of their own eyes; as they watched the WTC buildings drop perfectly into their own footprints at the speed of free fall, in violation of the Laws of Physics; including Building 7 *'pulled'* 8 hours later; pulverizing 'records' of the CIA, IRS, SEC and Secret Service?

"The covert operators that I ran with would blow up a 747 with 300 people to kill one person. They are total sociopaths with no conscience whatsoever."- Gene Wheaton

What American HS would have ever imagined that they would be so inconsolably distressed by the inexplicably dramatic contrast between the supposedly 'unaided' military precision of the first attack on *'The Great Satan'* by the international banking cabal's *useful idiot* mentally ill SOA, and the utterly moronic incompetent failure of all the SOA's supposed solo 'attempted attacks' since? What American HS would have ever imagined that their *'inalienable'* Civil Liberties, originally guaranteed to *'not be infringed'* by the Liberty honoring HS Sons of Great Britain's brilliantly crafted Bill of Rights; which so many have died to preserve; would be surrendered to The State without so much as a whimper of defiance from over *300 million* once earnest lovers of Liberty who'd now lost their courage to cowardice because *3000* of them were murdered? *Perspective?* What American HS would have ever imagined the historically complicit enemies of our species' preciously unique American experiment in Liberty gleefully laughing in their mosques and banks over how easy it was, even with failed 'attempted attacks', to maintain the 'faithful's fears, leading to the murdering of millions and profits in the trillions, leaving our entire planet again enslaved?

"According to some estimates, we cannot track $2.3 trillion in transactions."- Donald Rumsfeld

What American HS would have ever imagined what useful idiots the fearful BBA 'faithful' still remain for the *Evil Apes of Avarice* who manipulate them to achieve their own rapacious goals?

"The drive of the Rockefellers and their allies is to create a one-world government combining supercapitalism and Communism under the same tent, all under their control... Do I mean conspiracy? Yes, I do. I am convinced there is such a plot, international in scope, generations old in planning, and incredibly evil in intent."
-Larry P. McDonald- Assassinated- September 1st, 1983

NOTES

THE FED

"The few who understand the system will either be so interested in its profits, or be so dependent upon its favours, that there will be no opposition from that class. While on the other hand; the great body of people, mentally incapable of comprehending the tremendous advantage that capital derives from the system, will bear its burdens without complaint; and perhaps without even suspecting that the system is inimical to their interests."- The Rothschild Brothers

The 'Federal Reserve' is neither; it's not 'Federal', and there is no 'Reserve.' A group of *foreign* for-profit private banks, who create 'credit' out of thin air, then 'rent' it to us at compound interest, creating *endless debt*. We've allowed these Evil Apes of Avarice to codify themselves between us and our own 'currency'; with no public oversight; evading any audit; they control every aspect of our U.S. 'economy' by manipulating our 'credit supply' for their own profit. As bad as their created depressions and recessions are, even more insidious is their manipulation's *hidden tax* of inflation, which has devalued The Fed's 'Notes' by 98% since 1913, enslaving Americans by transferring their 'buying power' back to the 'issuers'. That combined with the 'rent' they charge us for their use about halves 'The People's actual 'wealth' each decade. So devious their lies, so complete their deception; they have ignorant BBAs fretting for fear of not being able to 'qualify' to become their *usury slaves*. Their Evil knows no ethnicity; 'The Fed' is a parasitic foreign corporation that profits at the extreme expense of every 'mortgaged' American child. Research *Amortization schedules* and *The Rule of 72*.

"Whoever controls the volume of money in any country is absolute master of all industry and commerce..."- James A. Garfield- Assassinated- July 2nd, 1881

In every way the antithesis of the letter and spirit of our Constitution's system of 'Checks and Balances', The Fed's 'Chairman' has become the unchecked, all powerful 'American despot' our Founders so feared, and hoped their brilliantly composed document would prevent at all costs. Looking eerily like one of Goodall's *'old men of the forest'*; ears and all; our collusive chattering Chimp congress fawned over Greenspan's every word; as 'faithful' before an oracle; as if only he *'divined'* an inerrant knowledge of our *'mystical disembodied'* economy. Wake up! Those foreign 'central bankers' are lying sociopaths who extort half of our work product every decade; pillaging the lives of our ignorant populous; enslaving them through deception; destroying the promise of our globally unique American Liberties with endless debt. Our HS Founders would be appalled.

"There's a thousand hacking at the branches of evil, to one who is striking at the root."- Thoreau

The Fed's private collection agency 'forbid' their usury slaves from owning 'offshore' accounts. Do your independent HS research; where are *they* 'domiciled'? Never 'hold' assets in your own name, but rather always in trusts. Use the same legal fictions they do; *'shells within shells within...'*

"It is well enough that the people of the nation do not understand our banking and money system, for if they did, I believe there would be a revolution before tomorrow morning."- Henry Ford

This sociopathic BBA child abuser financial construct; which enabled the foreclosure of millions of children's homes in the same year that the sociopaths who created the artificial crisis that caused all of those foreclosures *personally* profited in the billions; must be abolished. There can be no Peace without Justice, and no Justice without first annulling our 'central bankers' *extortion racket;* for the international banking cabal are behind all of the corruption in our governments. As in Arab nations, HSs have the Moral Courage to construct a public owned, interest free 'credit supply' in its place.

"Those who make peaceful revolution impossible will make violent revolution inevitable."
-John F. Kennedy- Assassinated- November 22nd, 1963

NOTES

TAXES

"The issue today is the same as it has been throughout all history, whether man shall be allowed to govern himself, or be ruled by a small elite." - Thomas Jefferson

How many Americans know that all of their Federal 'income' tax now goes directly to The Fed to barely 'service' our $17 trillion *debt;* which we could eliminate by abolishing The Fed and returning control of our 'currency' to congress as our Constitution mandated? How many know we now spend more each year murdering our siblings in their own lands, than we do caring for our own aged parents here in our land? Is that really the best use of our time here? Where will that lead us? Is America now run by sociopaths? Replacing the 74,000 page Federal tax code which you foolishly *'swear under penalty of perjury'* you're in compliance with; by signing your tax forms each year; HSs will institute a Flat National Sales Tax. Unrelated to 'income', and taxed only on what one 'spends', this FNST will be charged on all products and services, except food, water, medicine and housing; as to tax those would be abusive to children. All the revenue shall be allocated in four equal quarters to City, County, State and Federal, to keep the majority of the revenue under informed local control and utilization. As the only tax, it'll also include the only 'funding' allowed for all elections. It shall never be raised, and when those four jurisdictions have spent their allotments, none shall ever be allowed to 'credit' any more 'currency', forcing them to operate within their budgets. HSs will institute *single* 4 year terms of service, with rotating elections held every 2 years to limit the corrupted's ability to do harm before removal and forcing our servants to actually spend all of their time in office working for 'The People'; instead of their own re-elections.

"Experience hath shown, that even under the best forms of government, those entrusted with power have, in time, and by slow operation, perverted it into tyranny." - Thomas Jefferson

Ironically our treasonously complicit BBA politician's *immense wasted military spending* is a far greater threat to our National Security than any foreign army could ever be; as their recklessness may force our 'Dollar' to lose its exalted status as the world's most *reserved currency;* destabilized by too much debt from all their criminal foreign wars and repeatedly 'crediting' their sociopathic banker friends 'bailouts'. Delighting Russia and China; as it will end our ability to buy oil with our own U.S. currency; dropping our high American standard of living to a second class economy, as it did irreparably to Englands when their 'Pound Sterling' plunged in value after losing its reserve currency status. The best way to tame an ever more ravenous beast is starvation. War is when you're told who your enemy is, Revolution is when you figure that out for yourself and refuse to continue to fund American BBA State terrorism. Disobey!

"What country can preserve its Liberties if its rulers are not warned from time to time that their people preserve the spirit of resistance?" "I like a little rebellion now and then." - Thomas Jefferson

Along with eliminating the extreme burden of The Fed's hidden tax of inflation, and the compound interest they charge us to 'rent' our own currency, this Flat National Sales Tax will greatly enhance the freedom inducing mobility of every American. Or maybe our species isn't yet worthy of Liberty and Self-Determination? Perhaps BBAs need yet another few centuries with a boot on their throats? Ask all HS children to define slavery. If they're not slaves, what moral right does The State have to know their 'income' from their own labor? Can theft ever be justified by immoral petty Ape envy? Is expending billions of U.S. taxpayer Dollars annually 'defending' the 'Holy Land' leading by our HS Rational example, or funding an endless immoral cycle of deluded BBA 'faith' based conflict? If in all the Eons of time and space, this is the only life they'll ever know, don't they have a Moral Duty to refuse to allow any of their 'income' here to be channeled into murdering their siblings?

"To compel a man to furnish contributions of money for the propagation of opinions which he disbelieves and abhors is sinful and tyrannical."
-Thomas Jefferson-

NOTES

SELF-INFLICTED INJURY

"Find out just what any people will quietly submit to and you will have the exact measure of the injustice and wrong which will be imposed upon them." - Frederick Douglass

All recent history proves the way to bring down an 'Empire' is to attack it's economic base by duping them into protracted foreign wars. Despite their lies; their treasonous 'Patriot Act' hasn't stopped any useful idiot SOAs' further 'attacks'; because they'd already achieved that deception with the first one. And all the complicit American 'networks' tell the exact same five 'stories' each day. Who chooses those five 'stories' we're all going to be told? Whenever there's a *'big story'*, look for The Truth they're trying to distract you from. Daily railing about the Greeks' *'failure'* to pay their *'interest debt'*, yet never once mentioning Iceland's thriving economy since evicting the international banking cabal from their lands. While our deceived and abused Military run all over our planet murdering their siblings, benefiting no one except those rapacious financiers; our own 'central bankers' continue to devalue their 'Notes' with *'quantitative easing'*, and all their lackey politicians initiate ever more conflicts, to enslave us with ever more *debt* to their masters; while they divisively distract The People with their Gay Marriage nonsense. Their stores filled with non-food 'consumables', their hospitals filled with BBAs who ate them and don't know why they're sick. While they poison all our children with Fluoride and Ethylmercury, and fill their young minds with lies and nonsense; and their TSA unconstitutionally searches Americans to maintain their fears of, and focus on, non-existent *'terrorist threats'*; deceiving them into missing their very real foreign enemies here at home. And the BBA Media keeps chanting *'Be afraid, you need us; be afraid,...'*

"If you think of yourselves as helpless and ineffectual, it is certain that you will create a despotic government to be your master. The wise despot, therefore, maintains among his subjects a popular sense that they are helpless and ineffectual." - Frank Herbert

Not knowing The HS Truth and so putting their 'faith' in lies and nonsense, BBAs are simply incapable of making judicious life choices. While HSs quietly choose not to rebuild below sea level near a coast; nor below flood level in a river plain; nor live in 'tornado alley'; nor abuse their Ape body with alcohol, drugs or calories; nor join the State's criminally mandated *'exchanges'* that insure such asinine behavior. As our airline industry operates on very thin margins; if all Americans knew The HS Truth and simply refused to fly on a single given day of the week, their industry would collapse and our tyrannical State would nullify their treasonous 'Patriot Act's TSA searches; as The Fed won't tolerate that loss in profit. And when The Fed's 'Notes' soon collapse from the weight of their avarice, it'll make their 'depression' pale in comparison; so HSs now restrict their use to the absolute functional minimum by maintaining their own Optimum Health through diet and activity level; by producing their own best quality food and unfluoridated water and solar electricity; and by bartering for all other goods and services among fellow self-sufficient HSs. Known as *'the normalcy bias'*, which blinded the Jews in 1930s Germany; when food stops arriving at the clusters of dependency and all the disarmed Hivers there look inland; where all the food and personal firearms are; how do they imagine that is going to work out for them? Teach all HS children the BBA Media is a form of hypnosis used to manipulate BBA public opinion antagonistic to their HS Enlightenment. Rather than diminish BBAs' learned 'faith', 'race' or 'nation' based *'us'* and *'them'* bigotry; all war, even the Holocaust, only reinforces them. And its tragic legacy has been to label the HS Rationalists who now challenge those war inducing delusional 'faiths' as 'anti-Semitic', or even 'anti-American'. **The Truth is Not Hate and Delusion is Not Love.** And that lie has retarded our Moral Evolution for centuries and dishonors all those murdered innocent victims of 'faith'. In their honor we must stop all BBA 'military actions' proven to only reinforce those delusive 'faiths'; as BBAs cannot murder our way to Reason. Our only hope lies in universal childhood HS education.

"That men do not learn very much from the lessons of history is the most important of all the lessons of history."
-Aldous Huxley-

NOTES

EXTINCTION

We either honor The Truth, or we don't. We either always teach The Truth to our children, or we don't. Their very survival depends on knowing and freely speaking The Truth. That either matters, or it doesn't. We either follow the other ninety nine percent of all the species that came before us here into extinction, or we don't. We either allow our Ape testosterone and adrenalin and juvenile egomaniacal delusions of *'eternal'* grandeur to control our species' destiny, or we don't. Insisting Empirical Evidence based self-awareness derived empathetic Dignity and Reason guide it instead. All of the BBAs' 'problems'; from personal to international; results from them *not knowing what they are*. For the first ever generation in our entire Evolution, through The Scientific Method alone, Humane Sapients have confirmed with conclusive genetic certainty, The Truth of *what we are*, and *where we came from*, and therefore; *where we should be headed*. Only when every young child on Earth is taught that HS Truth, will its self-evident Peace assuring 'solutions' harmonize our planet.

"The further backward you can look, the further forward you are likely to see." - Winston Churchill

What do we entrust to prove what course will advance our Moral Evolution, if not 'The Method'? Only by the greatest percentage of our species knowing and all freely speaking and cooperatively applying The HS Truth will we be able to chart the most viable course for our species' thriving survival. To do anything less will ensure our eventual extinction. Currently HSs comprise just a tiny fraction of our population; and although thanks to the Internet, a much greater percentage is awakening to The HS Truth in this generation; we've seen that speaking that Truth openly today may yet be fatal, as BBAs think homicide an appropriate response to any question of their 'faith'. If they'd prefer their children raped or murdered to admitting The Truth, then we're in real trouble. Being afraid of the imagined once helped us survive; it may now hasten our extinction. HSs' acute ability to adapt to newly discovered Objective Truth, as all BBAs now must, is a bold testament to Darwin's brilliant insight; and we can no longer afford to humor the BBAs among us who refuse to honor that Objective Truth; so they must be called out and courageously challenged in every venue, as their deluded fantasy *'afterlife'* obsessed death cults will likely kill us all. All those deluded Apes have had thousands of years of hegemony, and by every measure possible, their delusional world-views have utterly failed. Thanks to them our planet has never been more dangerous or degraded. Never before in our species' entire Evolution have we had the leaders of divisively irreconcilable, fallacious 'Holy Book' driven nuclear Theocracies so antagonistically threatening our extinction. It will not be HSs' 'Method' that brings our species' demise; it'll be BBAs' 'faiths' alone to blame.

"Most of the greatest evils that man has inflicted upon man have come through people feeling quite certain about something which, in fact, was false." - Bertrand Russell

All evidence suggests that we're alone here and that no supernatural force will mourn our passing. So we better get our collective BBA act together and educate our delusional brothers and sisters before they pull our final curtain down with the heavy weight of their profound ignorance. If we HSs don't develop 100, 300, 500, 1000, 3000 and 5000 year plans of Empirical Evidence based Rational action for our species' thriving survival, nothing, and no one else is going to do it for us. As BBAs who 'believe' 'The Method' and 'faith' can coexist understand neither; we must choose! The Epoch of deluded Big Brained Apes is over; this is the Age of Enlightened Humane Sapients. HSs join Thomas Jefferson in vowing *"...eternal hostility against every form of tyranny over the mind of man."* For all HS adults, militancy in propagating *The One Proven Truth* is no vice, and moderation in challenging all the Big Brained Ape 'faithful's Bronze Age death cults is no virtue. What hideous nuclear Theocracy proximate extinction act is it going to take before you join us?

"Human history becomes more and more a race between education and catastrophe."
-H. G. Wells-

NOTES

FINAL THOUGHTS

I first began making notes for this 'Rationalist's Guidebook' in 1986 while flying around half the hemisphere for several years continually meeting, and working with, very bright entrepreneurs. Of whom I discovered as I got to know better, all possessed the intellectual capacity for perfectly rational thought in every aspect of their successful lives; except where their 'religious faiths' were concerned; where they uniquely seemed to be missing a part of their brains. So dramatic was this singular inconsistency in each one's thought process that I began to hypothesize that all 'religious faiths' must in fact be cults, into which their parents had 'indoctrinated' them at a very young age before they'd attained the organic brain maturity to make their own rational evaluations. Thereby instilling a potentially lifelong form of delusional mental illness. For even when presented with incontrovertible evidence proving The Truth of Evolution and the fallacy of their 'beliefs' they yet persisted in their 'faiths', while denying those proven Truths with bizarre rationalizations such as;

"...the Devil put those fossils there to deceive you and lead you away from God..."

At the beginning, I had asked the reader whether it would ever be possible to arrive at a True conclusion when starting from a False premise? I now ask the reader that same question again. The simple premise of Science is that everything on this planet, evolved on this planet, without any outside influence whatsoever. While the Bronze Age premise of religions is; that contrary to all our five evolved senses; some unknown, unseen, unheard, untouched, untasted and unsmelled external 'something' is instead the source of everything here. A delusion the 'faithful' assert with pride, while HSs know it's their attempt to assuage their own self-ignorance based fears of this *'mysterious'* life. Why is it so important that we get that premise right? What harm does their 'faith' do to our species? Anywhere on Earth, one can observe that all BBA conflict and suffering is rooted in the divisions created among us by their cult like 'faith' in false Origin Stories; instead of *The One Proven Truth*. And until we end their collective delusion, we will never end that incessant immoral conflict and suffering; or even begin to solve all of our other proximate extinction problems; *as 'The Method' proves that it is not possible to arrive at a True conclusion, when starting from a False premise.*

A creative imagination was 'selected for' in BBAs' struggle to survive; conflicts only arose when it was channeled through the False Premises of Bronze Age cults. When every BBA spends just an hour every day contemplating the Natural World through the prism of *The One True Premise*, the imagined deluded fantasy etiology of our species' *'us'* and *'them'* bigotries; and the Rational, Moral and Just solutions to all our species' conflicts and suffering will become self-evident to all. For that begins with the self-awareness of *what they are,* and *where they came from*, and *how they got here,* as that HS *perspective* eliminates fears of the imagined. And early childhood education is crucial, as experience proves that once 'indoctrinated' even seemingly rational adults have great difficulty 'curing' themselves of their childhood myth infections, no matter what evidence is later presented.

The most effective 'cure' begins with a Socratic progression. First get agreement on the validity of the culturally neutral Scientific Method. Secondly, agreement that evidence based Objective Truth is derived from its application. Third, that all the world's religions are equally contrary to that evidence based Objective Truth. Fourth, agreement to set aside all religions and the conflicts they create after thousands of years of suffering and try something new? And if the new Age of HS Enlightenment doesn't work out; the BBAs can always go back to endless war and suffering! Educating to reduce and eventually eliminate all BBA conflict and suffering is HSs' most noble pursuit. It may take generations or even centuries, but if we survive, only through education may every BBA one day become *subjectively* aware of *The Objective Truth.* But until they all do, our species will never see lasting Peace on this planet. For our children's children, will you join us?

"Be ashamed to die until you have won some victory for humanity."
-Horace Mann-

FINAL THOUGHTS

THE HUMANE SAPIENT ENLIGHTENMENT FOUNDATION · THE TRUTH

DISCOVER THE OBJECTIVE TRUTH

Darwin delayed publishing for two decades, and may never have published in Victorian England had he not learned that Wallace had independently discovered the same Objective Truth. Darwin thought deeply about it, and he well knew what it meant. His HS *perspective* changed everything. You've now had time to think about it. Do you well know what it means? Big Brained Apes' entire global culture is structured on Bronze Age lies and nonsense. Much of what they 'believe' is false. You must have the Courage to Trust yourself to discover The Objective Truth, and when *you know you have,* share it with all those worthy of it, and together make this world a better place for all our children. Always act in their interest, focusing on their future, never on our darkly ignorant past.

"Until we have the courage to recognize cruelty for what it is; whether its victim is human or animal; we cannot expect things to be much better in this world. We cannot have peace among men whose hearts delight in killing any living creature. By every act that glorifies or even tolerates such moronic delight in killing, we set back the progress of humanity." - Rachel Carson

Honor The Objective Truth; it is the only genuine Moral Authority in this brief singular life.

"The word 'God' for me is nothing more than the expression and product of human weaknesses, the Bible a collection of honorable but still primitive legends, which are nevertheless pretty childish. For me the Jewish religion, like all other religions, is an incarnation of the most childish superstitions." "A man's ethical behavior should be based effectually on sympathy, education, and social ties; no religious basis is necessary. Man would indeed be in a poor way if he had to be restrained by fear of punishment and hope of reward after death. In this sense I have never looked upon ease and happiness as ends in themselves; such an ethical basis I call more proper for a herd of swine. The ideals which have lighted me on my way and time and time again given me new courage to face life cheerfully, have been Truth, Goodness and Beauty." "It was of course a lie what you read about my religious convictions, a lie which is being systematically repeated. I do not believe in a personal God and I have never denied this, but have expressed it clearly. If something is in me which can be called religious then it is the unbounded admiration for the structure of the world so far as our Science can reveal it." "The further the spiritual evolution of mankind advances, the more certain it seems to me that the path to genuine religiosity does not lie through the fear of life, and the fear of death, and blind faith, but through striving after Rational Knowledge." "A human being is part of the whole, called by us the 'Universe', a part limited in time and space. He experiences himself, his thoughts and feelings as something separate from the rest; a kind of optical delusion of his consciousness. This delusion is a kind of prison for us, restricting us to our personal desires and to affection for a few persons nearest to us. Our task must be to free ourselves from this prison by widening our circle of compassion to embrace all living creatures and the whole of Nature in its beauty. Nobody is able to achieve this completely, but the striving for such an achievement is in itself a part of the liberation, and a foundation for inner security."

"Concepts that have proven useful in ordering things easily achieve such authority over us that we forget their earthly origins and accept them as unalterable givens. The path of Scientific progress is often made impassable for a long time by such errors. Therefore it is by no means an idle game if we then become practiced in analyzing long held and commonplace concepts and showing the circumstances on which their justification and usefulness depend, and how they have grown up, individually, out of the givens of experience. Thus their excessive authority will be broken." "We cannot solve problems by using the same kind of thinking we used when we created them." "A new type of thinking is essential if man is to survive and move toward higher levels." "I prefer an attitude of humility corresponding to the weakness of our intellectual understanding of Nature and of our own being."
-Albert Einstein-

Feed the Hungry; Educate the Ignorant; Preserve the Environment; Colonize Space

FINAL THOUGHTS

AWAKE, RISE, EDUCATE AND ACT!

"All mankind is divided into three classes: those that are immovable, those that are movable, and those that move." "Instead of cursing the darkness, light a candle."- Benjamin Franklin

After three decades of adult education experience, the author can affirm that if one spoon feeds information to adult Big Brained Apes, no matter how relevant or profound, it goes in one ear and out the other. But if one motivates them to *independently research and confirm* that information for themselves, it becomes a part of their own life experience that they pass on to their children. Awake and Rise my fellow Big Brained Ape brothers and sisters, courageously seek and confirm The Humane Sapient Truth for yourself and your children and together we will change this world!

"The teacher who is indeed wise does not bid you to enter the house of his wisdom; but rather leads you to the threshold of your mind."- Khalil Gibran

America's hidden tax of inflation has historically averaged over 3%; while 'credit' rates to 'rent' our own currency, have averaged over 4%. Applying The Rule of 72, one can see that the 'cost' you pay for The Fed's Notes 'rented' to you by our foreign 'central bankers' is about half their buying power every decade, compounding throughout your entire life. Their deception is the 'cost' of everything else keeps going up; in reality the value of their Notes keeps going down. The simple Truth is they're stealing from you. And as you 'refinance' every five years on average, you're farmed for your work product on endless Amortization schedule treadmills by The Fed. A prevalent regret expressed by the dying is the wish that they hadn't spent so much time working. You've always had a sense that our 'economy' wasn't Just; but never imagined you'd been enslaved. Can you even imagine the freedom inducing mobility of having twice the buying power that you now have every decade, compounding throughout your entire life without changing anything you do except nullifying our central bankers' extortion racket, and returning control of our 'currency' to congress as The Constitution mandated? Both the Arab Saracen Empire and the Mandarin Chinese issued their own interest free currency; and anthropologists and historians consider both societies to be among the very best flowerings of our species' Artistic, Cultural, Scientific, and Economic expansion in our entire global Evolution. America once had that great promise too; and our whole species still looks to us to lead by example as we did with our Moon Shot; but now we can't even afford to feed and house all our own children. The difference between America and those two great societies being our 'central bankers' extortion. Criminal Ape sociopaths are now running our planet; it's long past time we took it back from them.

"Only small secrets need to be protected. Big ones are kept secret by public incredulity."- M. McLuhan

If we are to lead our species' Moral Evolution, we must now follow Iceland's courageous example. This isn't rocket Science, it's seventh grade math; they just want you to 'believe' that our *'mystical disembodied'* economy is rocket Science; that only *they* can *'divine.'* It's actually stupid simple once one knows The HS Truth. Problem is; you've been told lies for your entire life, by all those you loved and trusted to tell you The Truth. Compounded by the fact that they did love you and 'believed' they were telling you The Truth. And that is a very difficult childhood myth infection to 'cure' oneself of; but have the Courage to Trust yourself to confirm The HS Truth and join us! To protect their entrenched archaic treachery, the current sociopathic Big Brained Ape hegemony will continue to jail and murder Humane Sapients, but either action only proves The HS Truth, and exposes those despotic Evil Apes of Avarice for what they are; *which is their greatest fear;*
The Humane Sapient Enlightenment.

"What good fortune for those in power that the people do not think." "It also gives us a very special secret pleasure to see how unaware the people around us are of what is really happening to them."
-Adolph Hitler-

FINAL THOUGHTS

EVIL APES OF AVARICE

"Of all the enemies to public Liberty war is, perhaps, the most to be dreaded, because it comprises and develops the germ of every other. War is the parent of armies; from these proceed debts and taxes; and armies, and debts, and taxes are the known instruments for bringing the many under the domination of the few... inequality of fortunes, and the opportunities for fraud, all growing out of a state of war... no nation could preserve its freedom in the midst of continual warfare." "A standing military force, with an overgrown Executive will not long be safe companions to Liberty. The means of defense against foreign danger, have been always the instruments of tyranny at home."
-James Madison- Considered 'The Father' of The Constitution

"The bold effort the present central bank had made to control the government; are but premonitions of the fate that await the American people should they be deluded into a perpetuation of this institution, or the establishment of another like it." "Gentlemen, I have had men watching you for a long time, and I am convinced that you have used the funds of the bank to speculate on the breadstuffs of the country. When you won, you divided the profits amongst you, and when you lost, you charged it to the bank (taxpayers)... you are a den of vipers and thieves!" "If congress has the right under the Constitution to issue paper money, it was given to be used by themselves, not to be delegated to individuals or corporations (banks)." "I am one of those who do not believe that a national debt is a national blessing, but rather a curse to a Republic; in as much as it is calculated to raise around the administration a moneyed aristocracy dangerous to the Liberties of the country." "The bank... is trying to kill me, but I will kill it!"
-Andrew Jackson- Assassination attempted- January 30th, 1835

"The money powers prey upon the nation in times of peace and conspire against it in times of adversity. It is more despotic than a monarchy, more insolent than autocracy, and more selfish than bureaucracy. It denounces as public enemies, all who question its methods or throw light upon its crimes. I have two great enemies, the Southern Army in front of me and the bankers in the rear. Of the two, the one at my rear is my greatest foe." "As a result of war, corporations (banks) have been enthroned and an era of corruption in high places will follow, and the money powers of the country will endeavor to prolong its reign by working upon the prejudices of the people until the wealth is aggregated in the hands of a few, and the Republic is destroyed."
-Abraham Lincoln- Assassinated- April 15th, 1865

"Now more than ever before, the people are responsible for the character of their congress. If that body be ignorant, reckless and corrupt, it is because the people tolerate ignorance, recklessness and corruption. If it be intelligent, brave and pure, it is because the people demand these high qualities to represent them in the national legislature. If the next centennial does not find us a great nation, it will be because those who represent the enterprise, the culture, and the morality of the nation do not aid in controlling the political forces."
-James A. Garfield- Assassinated- July 2nd, 1881

"What kind of Peace do I mean, and what kind of a Peace do we seek? Not a 'Pax Americana' enforced on the world by American weapons of war. Not the Peace of the grave, or the security of the slave. I am talking about genuine Peace. The kind of Peace that makes life on Earth worth living. The kind that allows men and nations to grow, and to hope, and to build a better life for their children. Not merely Peace for Americans, but Peace for all men and women. Not merely Peace in our time, but Peace in all time... The United States, as the world knows, will never start a war. We do not want a war, we do not now expect a war. This generation of Americans has already had enough; more than enough; of war and hate and oppression. We shall be prepared if others wish it, we shall be alert to try to stop it, but we shall also do our part to build a world of Peace, where the weak are safe, and the strong are just. We are not helpless before that task, or hopeless of it's success. Confident and unafraid, we must labor on, not toward a strategy of annihilation, but toward a strategy of Peace."
-John F. Kennedy- Assassinated- November 22nd, 1963

FINAL THOUGHTS

WHAT ARE YOU PREPARED TO DO?

"There is no substitute for a militant freedom. The only alternative is submission and slavery." - C. Coolidge

Take a long hard look in the mirror America. What have we allowed those Evil Apes of Avarice to do to our once great promise? Remember when Soviet and Chinese dissidents used to seek asylum here, instead of the other way around? Maybe BBAs are not yet worthy of Liberty and Self-Determination? Perhaps our species needs yet another few centuries with a boot on their throats? The reason the Evil Apes of Avarice put a camera on every corner, and illegally monitor The People's private electronic communications isn't external *'terrorism'*. It's because they themselves are terrified of *'us'* nullifying what they view in their Ape delusions as their *'power'*; by educating all those worthy of The HS Truth of their criminal Ape lust for it. All HSs have then a Moral Duty to maintain those Evil Apes' fears; by shining the hot light of Truth on their murderous avarice. Can't you just smell their fear? *They are not Americans.* Scoffing at The Constitution as they dishonor our historical American Moral Courage to Trust all others with Liberty and Self-Determination; those disgraced cowards in D.C. only represent themselves and their sociopathic 'central banking' masters, who seek only our species' total financial enslavement. Do not be silenced! *Exposure is their greatest fear;* as they can only continue to pursue their murderous avarice in the darkness of universal BBA ignorance. Courageously turn up the light! Educate others! And as they are using your own work product to oppress you; stop giving it to them!

"Once a government is committed to the principle of silencing the voice of opposition, it has only one way to go, and that is down the path of increasingly repressive measures, until it becomes a source of terror to all of its citizens and creates a country where everyone lives in fear."
"If even one American- who has done nothing wrong- is forced by fear to close his mind and shut his mouth- than all Americans are in peril." - Harry S. Truman

All BBA inflicted suffering on this planet is rooted in delusional 'faith'. Delusional 'faith' in Bronze Age Harry Potter novel 'Holy Books'; delusional 'faith' in Nationalism; and delusional 'faith' in Profit Worship. This is our children's planet the Evil Apes of Avarice are destroying. It is our children they lie to and send off to murder others' children for their own rapacious interests. If we HSs don't stop all those 'faithful' BBAs from their homicidal Simian rampages; they'll never stop themselves. Now all that stands between a better world for all our children's children is *your* apathy or action. Whose planet is this? With whom do you stand? The day for that decision is here; *make it!* Reach out and spread The HS Truth to all you can in our deceived and abused Military and Police; insisting they all honor their oaths to *'Defend our Constitution'*; not criminal orders from disgraced D.C. cowards. War is when you are told who your enemy is; Revolution is when you figure that out for yourself.

"If ever time should come when vain and aspiring men shall possess the highest seats in government, our country will stand in need of our experienced patriots to prevent its ruin." - Samuel Adams

Our 'central bankers' will never nullify themselves, those sociopathic Apes would sooner murder millions to distract *'us'* from their profits, than to allow the advance of our species' Moral Evolution. Viewing this world through the prism of your new HS *perspective;* you'll see that they're the true barrier to our species' HS Enlightenment, not our useful idiot 'faithful' siblings they manipulate to do their bidding. So educate all those worthy of The HS Truth to strike at the root of their species' enslaving evil. Do not leave it for our progeny; as they'll have their hands full dealing with our self-inflicted severe climate change. But mostly, have Courage, and with HS Geologic vision, Trust that we will one day be rid of their Evil.

"Truth alone will endure, all the rest will be swept away before the tide of time. I must continue to bear testimony to Truth even if I am forsaken by all. Mine may today be a voice in the wilderness; but it will be heard when all other voices are silenced, if it is the voice of Truth."
-Mohandas Gandhi- Assassinated- January 30th, 1948

FINAL THOUGHTS

'CENTRAL BANKING' FOR BBA 'CHILDREN'

"The real Truth of the matter is, as you and I know, that a financial element in the large centers has owned the government of the U.S. since the days of Andrew Jackson." - Franklin D. Roosevelt

A long long time ago, in a land far far away, a very *'select'* group of Apes owed all the bananas. And if any of the many *'plain'* Apes living there in that land with them needed a banana, those very select Apes would gladly *'rent'* one to them. Because that very select group of Apes also had a *'magic machine'* with which they could *'create'* as many bananas as they wanted out of the clear thin air, with no limitations of any kind, and at absolutely no expense to themselves! But all the many plain Apes that rented one of those select Apes' bananas, all had to *'pay back'* every banana they rented, with two bananas of their own! And every single banana those plain Apes rented, through no fault of their own, would *'shrink'* by half every twenty years, no matter where those plain Apes kept it! But as those many plain Apes didn't have a *'magic machine'* to create bananas out of the clear thin air for themselves, the only way they could pay back those select Apes all of their shrinking bananas, was to rent even more new shrinking bananas from them! This process went on and on and on, as no matter what new lands the many plain Apes traveled to, the select Apes always followed, and insisted on repeating the process; while also always inciting costly conflicts among all the plain Apes, because for every one shrinking banana the plain Apes had to rent to pay for those costly conflicts, they had to pay back the select Apes two! Until finally, all the many *honest* plain Apes, and even all their many *honest* plain Ape children and grandchildren, just simply could not pay back those select Apes all of their shrinking bananas, no matter how hard they all *honestly* tried!

But then one day; despite all the many lies from the select Apes' *'Holy'* Ape and *'State'* Ape friends who were all *very dependent* upon the select Apes' *'magic machine'*; a small number of the plain Apes *came to realize* that those select Apes weren't really select after all, but just fellow plain Apes just like them! *Having no 'magical' powers!* And that their *'magic machine'* wasn't really *'magic'* after all! And that the *honest* plain Apes could run that machine for themselves! And figure out just how much those created bananas were really worth, without having to shrink all of their rented bananas, or unfairly asking for two to be paid back, instead of the one they originally rented! Simply said; *they realized* those select Apes were *stealing* from all of the plain Apes! *And not just a little!* As added to the 'rent' they charged them to use their bananas; they were stealing back *half* of all of their bananas every decade! And then that small number of *self-aware honest* plain Apes began to *educate* all the rest of their fellow *honest* plain Apes as to The Truth about those lying thieving select Apes and all of their lying Ape friends!

"That which can be destroyed by The Truth should be." - Patricia C. Hodgell

And that didn't make the sky fall!!! As all those lying thieving select Apes and all of their lying Ape friends had all claimed would happen; *if The Truth was finally told!* As every day the Sun came up, and the rains fell down, and all of the plants and animals grew, and all the many *honest* plain Apes then had all the bananas that any of them ever needed! *And for the very first time ever* there was True Justice and Liberty and Equity and Prosperity and lasting Peace throughout all of the many *honest* plain Apes' lands!

"Truth never damages a cause that is Just." - Mohandas Gandhi

Well children? Should the *honest* plain Apes have just continued to allow those lying thieving select Apes and all of their lying Ape friends to continue to *steal* from them by 'renting' them their shrinking bananas? Or should the *Truth honoring* plain Apes have simply started running the banana machine for themselves and their *honest* plain Ape children and grandchildren happily ever after? After all children, what *honest* kids would allow *lying* kids to continue to *steal* from them once they knew of it? **Would any of you?**

"I freed a thousand slaves. I could have freed a thousand more, if only they knew they were slaves."
-Harriet Tubman-

FINAL THOUGHTS

TEACH YOUR CHILDREN WELL

Their HS *perspective* is their greatest strength; teach your HS children to wield it wisely.

"Never be afraid to raise your voice for honesty and Truth and compassion against injustice and lying and greed. If people all over the world... would do this, it would change the Earth." - W. Faulkner

Always remember, our species' collective knowledge is estimated to now double every decade. Every construct of the Big Brained Apes that came here before you were all made up by Apes that were *no smarter than you,* while having access to *far less accurate information than you.* Potentially fatal flaws in our species' character are hubris, credulity and unwarranted trust. Never unquestioningly accept any of their constructs as being correct, much less the best possible option for our species. *Question everything* you've ever been told or shown. *'Cui bono?'* Do not be deceived! The largest room in this world is the room for improvement.

"We are now at a point where we must educate our children in what no one knew yesterday and prepare our schools for what no one knows yet." "Never doubt that a small group of thoughtful, committed citizens can change the world; indeed, it is the only thing that ever has." - Margaret Mead

One may structure their life on Empirical Evidence based Objective Truth; or on Bronze Age lies and nonsense. Which of those pursuits will ensure a more secure future for their grandchildren? A BBA's 'faith' is no longer just a choice for themselves alone. With our rapidly changing global atmosphere; that choice now impacts their grandchildren's lives more than their own. With their thriving survival now in the balance, there exists only one moral option for all BBAs to choose. For far too long we've allowed the Scientifically illiterate Profit Worshiping BBAs among us to betray our grandchildren by delaying our response to climate change. As the 'faithful' all want to 'believe' that the lying obese radio talk show climate change 'deniers'; who can't even connect the obvious causation of their own conspicuous consumption; are somehow qualified to determine the effects of global scale conspicuous consumption better than 'The Method'. All of our grandchildren will revile them for costing us preciously unrecoverable time; as their suffering will be magnified if the myth infected BBAs maintain their 'faith' in those that worship profit driven lies over Truth.

"Men argue. Nature acts." - Voltaire

We're changing Earth's atmosphere so dramatically, BBAs must now *adapt or perish.* The first step towards adapting is to eliminate entrenched archaic financial parasites. We simply can't afford them any longer; we just have too much *adult* work to do. They've *'had their way with us'* since the 1400s in Europe, but now they've got to go. Cash flow is all that limits our innovation and progress; those parasites have repressed our species' progress by half every decade since. It's long past time we put our species' financial house in order. *Our very survival now depends on it.* If we do not now nullify The Fed, our extinction is assured; it's just a matter of time. The international banking cabal are not *'too big to fail',* but rather too big to not be 'removed' from imperiling our species' thriving survival. Profiting several compound interest points on each of their 'Notes' that they 'rent' into our 'economy' every year, rather than progress, their incentive is to have us *waste* as many as possible on expensive futilities; no wonder America's been warring endlessly your entire lifetime! We all have then a Moral Duty to our children to disobey their criminal 'laws' by refusing to continue to fund their murderous sociopathic Ape avarice. Can Americans even imagine a world where their Congress actually spends all of its time 'regulating' their 'currency'; as The Constitution mandated; instead of trying to 'regulate' their bedrooms? Where did they even find all of those undignified moral cowards? They too must go.

"There is no power for change greater than a child discovering what he or she cares about."
-Seymour Simon-

FINAL THOUGHTS

LISTEN LESS AND THINK MORE! LLTM!

"It is easier to judge the mind of a man by his questions rather than his answers." - Gaston de Levis

Have the Courage to Trust yourself to *independently research and confirm* The HS Truth.

"The most important thing is to never stop questioning. Curiosity has its own reason for existing." "I have no special talents. I am only passionately curious." "I have only two rules which I regard as principles of conduct. The first is: Have no rules. The second is: Be independent of the opinion of others." "Great spirits have always encountered violent opposition from mediocre minds. The latter cannot understand when a man does not thoughtlessly submit to hereditary prejudices but honestly and courageously uses his intelligence." "What a betrayal of man's Dignity. He uses the highest gift, his mind, only ten percent, and his emotions and instinct ninety percent." "Nationalism is an infantile disease; the measles of mankind." "Force always attracts men of low morality." "Peace cannot be kept by force; it can only be achieved by understanding." "The world is a dangerous place to live; not because of the people who are evil, but because of those who look on and do nothing." "Example is not another way to teach, it is the only way to teach." - Albert Einstein

Moral Courage derives from The Humane Sapient Truth *perspective;* derived from Geologic vision. Henceforth Humane Sapients will put these four questions to every BBA society 'authority' candidate. All of their answers to which will either garner our unwavering support, or our unrelenting opposition: What are we? Where did we come from? How did we get here? How will that inspire all of your efforts towards the advancement of our species' Enlightened Self-Awareness and resulting Moral Evolution?

"Some men see things as they are and ask why? I dream of things that never were and ask why not?" "Those who dare to fail miserably can achieve greatly." "The purpose of life is to contribute in some way to making things better." "Too much and too long, we seem to have surrendered community excellence and community values in the mere accumulation of material things." "Too often we honor swagger and bluster and wielders of force; too often we excuse those who are willing to build their own lives on the shattered dreams of others." "Every time we turn our heads the other way when we see the law flouted, when we tolerate what we know to be wrong, when we close our eyes and ears to the corrupt because we are too busy or too frightened, when we fail to speak up and speak out, we strike a blow against freedom and decency and justice." "Few are willing to brave the disapproval of their peers, the censure of their colleagues, the wrath of their society. Moral Courage is a rarer commodity than bravery in battle or great intelligence. Yet it is the one essential, vital quality for those who seek to change a world that yields most painfully to change." "It is from numberless diverse acts of courage and belief that human history is shaped. Each time a person stands up for an ideal, or acts to improve the lot of others, or strikes out against injustice, they send forth a tiny ripple of hope, and crossing each other from a million different centers of energy and daring those ripples build a current which can sweep down the mightiest walls of oppression and resistance." "This world demands the qualities of youth; not a time of life but a state of mind, a temper of the will, a quality of the imagination, a predominance of Courage over timidity, of the appetite for adventure over the life of ease." "The responsibility of our time is nothing less than a revolution. A revolution that would be peaceful if we are wise enough; humane if we care enough; successful if we are fortunate enough. But a revolution will come whether we will it or not. We can affect it's character, we cannot alter it's inevitability." - Robert F. Kennedy- Assassinated- June 5th, 1968

The Moral Courage of the HS *perspective;* that additional 80 IQ points *'vision'* thing; *got it?* Then share it with all those worthy of it, and together 'remove' all the criminal Ape sociopaths blocking our species' HS Enlightenment and resulting Moral Evolution as they profit by all our self-ignorance based conflicts.

"The Duty of Youth is to challenge corruption."
-Kurt Cobain-

"The best way to predict the future is to invent it." - Alan Kay

History has proven the 4% of our species that are born conscienceless sociopaths are always attracted to centralized 'authority'; the more 'centralized', the more attracted; often with an extinction risk to all the rest of us. Consequently, our planet is now predominately run by the sociopaths. Some visible and known to us and some not; but always threatening our species' extinction with their murderous avarice; yet claiming they're acting with only the purest of intentions; while denying others their Inherent Right to Liberty and Self-Determination. Stop believing their lies and funding their global terrorism! *Learn the known signs of Sociopathy!* As we cannot alter their Ape biology; the HS *perspective* realizes that to prevent our extinction and continue to advance our Moral Evolution; we must now alter our species' global culture instead. We all have then a Moral Duty to all children to commit all our lives to disarming and disbanding every nation's armed forces; melting their large offensive weapons down into plowshares and green renewable energy production; and distributing their small personal defensive firearms among all of their able citizenry; by following the courageous example of the virtually crime free Swiss culture. As a universally armed able populous will defend their own righteous 'government', or change a corrupt one by themselves; and history has proven that a standing 'State' army is always used to enable its own despotic treachery. As the last thing that any free citizen of any moral society needs is a large number of heavily armed non-self-aware Apes running around among them immorally enforcing their own deluded worldviews about *'lifestyle'* choices. Something that would be self-evident to them; if they simply knew The HS Truth about *what they are,* and *where they came from,* and *how they got here;* instead of all the myth infected 'faithful's Bronze Age lies and nonsense that they've been taught all of their lives. So HSs will also reduce every deceived and abused Police force to a streamlined team of HS professionals solely charged with catching and medically 'certifying' all of the violent criminal sociopaths among us towards their permanent removal to 'the islands'. And sending the despotic sociopaths that currently run our own corrupt government and the international banking cabal towards our species' financial enslavement to the empty Ape Zoos. Finally changing our species' sick global culture of endless war to one of lasting Peace. So we can start working on our *real* self-inflicted proximate extinction problems that may yet kill us all.

"No matter how paranoid or conspiracy-minded you are, what the government is actually doing is worse than you imagine." - William Blum

Educate All Those You Meet Who Are Worthy of The Humane Sapient Truth

Whose country is this?   http://stormcloudsgathering.com/the-declaration-of-natural-rights

Are you funding your own oppression?   http://www.youtube.com/watch?v=VEzRcsfoVA4

Will your great grandchildren starve?   http://www.youtube.com/watch?v=cA3IJNQbKW0

Too sensible for D.C. funding?   https://www.youtube.com/watch?v=SNMFKKyFU60#t=273

Will our own Ape hubris be our demise?   http://www.youtube.com/watch?v=ypXu9JMGFuc

Should Apes make bombs or clean energy?   https://www.youtube.com/watch?v=g6wQP2qaaEk

Is this Science or Politics?   http://www.youtube.com/watch?v=KiT7Y233404&feature=youtu.be

Are clueless Apes poisoning all the rest of us?   http://www.youtube.com/watch?v=ViNNIwmzTzI

"We suppress our genius only because we haven't yet figured out how to manage a population of educated men and women. The solution... is simple and glorious. Let them manage themselves."
-John Taylor Gatto-

Made in the USA
Charleston, SC
02 July 2014